MOTHS
THAT DRINK
ELEPHANTS'
TEARS

About the author

Matt Walker lives in London and is a writer and senior editor with *New Scientist*, the world's leading weekly science and technology magazine.

Born in 1973, Matt first tried to tell his parents the difference between a leopard and cheetah when he was still a toddler. His interest in the natural world continued, he was educated at the University of Leeds and Imperial College, London, where he gained a first-class bachelor's degree in Zoology and a master's degree in Science Communication.

Matt joined *New Scientist* in 1999 as a reporter, winning a national magazine award for best young journalist in the same year. He then moved into editing and for the past five years he has overseen the award-winning News and Technology sections of *New Scientist*, shaping the magazine's international coverage of science and technology issues. He is also a contributor and commentator on science and technology issues on local, national and international radio and television programmes, and gives talks at academic institutions, other media organisations and university courses on science journalism.

However, zoology remains his first love, and as well as specialising in breaking and reporting stories on new findings in zoological research at *New Scientist*, Matt considers reading and writing about the latest animal discoveries both a passion and a pastime.

MOTHS THAT DRINK ELEPHANTS' TEARS

And other zoological curiosities

Matt Walker

PORTRAIT

Visit the Portrait website!

PORTRAIT Portrait publishes a wide range of non-fiction, including biography, history, science, music, popular culture and sport.

If you want to:
- read descriptions of our popular titles
- buy our books over the Internet
- take advantage of our special offers
- enter our monthly competition
- learn more about your favourite Portrait authors

VISIT OUR WEBSITE AT: www.portraitbooks.co.uk

This book is for Morgan, Marnie and Matthew.
I hope you have as much enjoyment reading it one
day as I have had in writing it.

Acknowledgement

The curiosities reported in the book belong to the scientists and researchers whose inquisition, dedication and hard work led to their discovery and made this book possible.

Introduction

We have an endless fondness and fascination for animals. We keep pets, are captivated by the animals scurrying around our gardens and will spend large sums of money to indulge the simple pleasures of listening to tropical birds sing beautiful ballads, watching whales swim majestically through the oceans and following lions hunting big game on the savannahs of Africa.

It is that love of the natural world and an unavoidable curiosity to know more about the animals with which we share our lives that led me to start researching this book. It was also the realisation that much of our understanding of animals comes from the same, often anecdotal, sources or from similar reports of the behaviour of familiar species such as chimpanzees or big cats, which are both engaging and popular with zoologists and the public alike.

There is, however, so much more to life on this planet. There are animals that many of us have never heard of, and many species that exhibit little-known behaviours that are surprising and shocking. Others are simply strange, mysterious or so like ourselves in how they deal with the fundamental struggle to live and breed that their stories deserve to be told. A book that started out being an eclectic collection of animal oddities has ended up being a celebration of the animal kingdom.

Animals are what they are, and anthropomorphising about their intentions is always problematic. But it is salutary to realise just how alike animals and people can be. Monkeys like to get drunk, rats enjoy drugs, whereas birds go on dates as teenagers and then get divorced from their partners just as readily as we do.

Others are more peculiar. Some male beetles can impregnate

females they have never met, while virgin male butterflies make better lovers than males more experienced in the bedroom arts. Spiders chew off their genital organs to give themselves a better chance of successfully mating, while female fish fake orgasms to dupe males into thinking they have copulated. The female then swims off to find a better partner. The least attractive female sheep often mate with the most males, while male garter snakes have a propensity to become transvestites and dress up as females.

Yet more animals show an incredible duty to their newborn offspring. Killer whales and dolphins appear unique in that they and their calves never sleep for the first month of the offspring's life. This surprising ability ensures the calf can breathe and remain safe from predators, but staying awake for this long would kill most other species, including ourselves. Birds reaching old age suddenly put all their effort into becoming good fathers, seemingly aware that time is short to do the right thing by their young fledglings.

Such discoveries have been made only recently and illustrate the huge diversity of animal life. Yet many of these amazing behaviours, and much of what we understand about animals, remain hidden from the ordinary enthusiast, buried within scientific literature that can be difficult to access and read. For people who do make a career out of studying animals, new insights into the lives of birds, mammals, and reptiles and amphibians continue to bring joy and fascination. But no one can hope to appreciate or understand the differing lifestyles and habits of potentially millions of species, and even most zoologists remain expert in only a relatively small field of research, such as animal biomechanics or physiology, and specialise in the study of a discrete group of animals such as rodents or primates. *Moths That Drink Elephants' tears* seeks to bring some of those insights discovered by zoologists to the wider reader.

The book is not meant to be an academic treatise on each and every animal mentioned, and each entry cannot do justice to the lifestyle of each species. I have also afforded myself a few luxuries. For example, there are entries on bacteria, which are technically not animals, but they too show some eclectic behaviour that I could not resist reporting. Species names have been included where possible, as there is much confusion about the exact identity of many animals which are known by different names by different people in different parts of the world. There is even disagreement among scientists over correct species names, so I have used the Linnean name reported by the scientists who conducted the research that forms the background for each entry.

Anecdotes about animals are both popular and widespread. Unfortunately, many are not wholly accurate, so the book reports only on animal facts gleaned from scientific literature, which includes research journals, academic meetings, professional societies and conservation organisations. The curiosities reported have been validated by expert research, experimentation and observation. Sources for each entry can be found at the back of the book. Many curiosities reported have also been made only recently, which serves to illustrate just how little we still know and understand about the animals around us. Because of that, and because science continually seeks to find out more about the natural world and how it works, we may well come to learn new details about many of the entries included.

Most of all, I hope you will find this collection of zoological curiosities as fascinating as I do. There is a mysterious giant ape roaming the African jungle, male racehorses tend to be right-handed, or legged, whereas monkeys will pay to look at images of other sexually attractive primates. Rats can learn the difference between Dutch and Japanese, elephants have over 30 different ways to talk to each other, whereas Dracula ants suck the blood of their young, and seagulls affected by pollution regularly fall over. And if that does not prick your interest, then perhaps your eyes will be hopelessly drawn towards the next entry, as they move to discover what is the only invertebrate capable of getting an erection.

The mating game

∾

Birds can act like lovelorn teenagers. At around one year old, **barnacle geese** (*Branta leucopsis*) begin going on dates, sampling several potential mates by engaging in so-called 'trial liaisons', which may last from a few days to several weeks. They will keep dating until they settle on a final mate for life at around the age of two or three. Both males and females that find their final partner early in life also raise a family sooner than those that play the field.[1]

Animals look for compatible mates just as people do. **Pinyon jays** (*Gymnorhinus cyanocephalus*) that find a partner similar in weight, age and bill length produce more offspring, whereas similarly sized **barnacle geese** (*Branta leucopsis*) reproduce better than pairs compromising a large and small individual. **Great tits** (*Parus major*) look for partners that have a similar desire in exploring the environment. Two fast explorers, and two slow explorers, produce more offspring than a pair of birds comprising a slow and fast explorer, whereas **willow ptarmigan** (*Lagopus lagopus*) are more reproductively successful when they pair up with another bird of a similar social standing.[2]

Animals also divorce each other just as readily as people do. **Oyster-catchers** (*Haematopus ostralegus*) are a monogamous species where the same male and female birds come together each year to raise a chick. But each year, just under 8 per cent of oystercatcher couples split up. In the majority of these cases, the females instigate the divorce, deserting partners who have not been particularly successful at breeding in the past for a better quality of life with another bird nearby. Some females walk out on males who routinely knock eggs off their cliff perches, or fail to bring home enough food. And while some divorcees remain on the shelf, most go on to have much greater reproductive success over their lifetime than they would have done had they remained in their dead-end relationship. Divorce rates in birds range from 0 per cent in the utterly faithful **wandering albatross** (*Diomedea exulans*) to 99 per cent in the rather more fickle **greater flamingo** (*Phoenicopterus ruber roseus*).[3]

Around 3 per cent of all bird species practise polyandrous breeding, where three or more adults will live together in a nest and bring up young. Up to five adults have been known to share a nest this way. The practice is especially common in raptors such as **eagles** and **falcons**, and in over half such species, the nest is shared by unrelated birds.[4]

Male and female **Eurasian oystercatchers** (*Haematopus ostralegus*) particularly like to indulge in threesomes, but – like many three-way relationships – sometimes all the individuals get along, and sometimes they do not. When the birds, which are common on mud flats, form a threesome, it always comprises one male and two females. But the threesomes occur in two forms, aggressive and cooperative. In an aggressive threesome, each female will defend her own nest, while the male oystercatcher defends a territory that includes the nests of both females. The females lay eggs about two weeks apart, and proceed to attack each other frequently throughout the day. The male also has a favourite partner, contributing most of his parental care to eggs laid first by one mate, and often leaving the second nest unguarded. However, in a cooperative threesome, the two females share one nest, and both females will lay eggs together, laying each clutch about one day apart. The females also attempt to mate with each other frequently during the day, only slightly less often than they do with the male. The females also sit together and preen their feathers together. What's more the male has no favourite, and all three birds defend the nest together.[5]

The **blue-footed booby** (*Sula nebouxii*), a sea bird that usually nests on cliffs and rocky outcrops, continually points its beak towards the sky as part of its courtship display. The behaviour is thought to have evolved from the same movements the bird uses when it intends to take to the sky.[6]

A female **penduline tit** (*Remiz pendulinus*) attempts to conceal her eggs from her mate until she has laid the entire brood. The reason for the deception is that the female wants to ensure all her eggs are in place before she promptly abandons them to be cared for by the unsuspecting male.[7]

For some animals, sex can literally be a life-changing experience. When a male **seed beetle** (*Acanthoscelides obtectus*) has sex with a female, it inseminates her with chemicals that either make her die younger, or live longer. During mating, males of this species transfer different peptides and proteins to females. These can have beneficial effects, as some of the compounds appear to increase the number of eggs laid by a female, and females that have been mated tend to live longer than females that have not. They can also have a detrimental impact and be toxic to the female. However, males that tend to reproduce late on in life will inject females with more beneficial compounds, in essence extending her lifespan, than males that tend to reproduce early in life. By doing so, late-breeding males ensure that females will live longer, and hence be around to bear them more offspring, maximising the number of times their genes are passed down the generations.[8]

A similar thing also happens to bumblebees, except female **bees** (*Bombus terrestris*) that have been inseminated with sperm by multiple males die younger, being less likely to survive hibernation, than those that have successfully mated with only one male. It is not clear why, but it could be that multiple inseminations of sperm sap more energy from the queen bee, or that chemicals within the sperm of each male, which are designed to corrupt the sperm of a rival in the bee equivalent of a sperm war, may also physically harm the female.[9]

The most attractive male **field crickets** (*Teleogryllus commodus*) die young, the price they pay for putting so much effort into making chirping calls designed to attract potential mates.[10]

A male **flour beetle** (*Tribolium castaneum*) can mate and impregnate a female he has never met. No other animal is known to have sex by proxy in this way. Flour beetles lead promiscuous lives, and multiple males often mate with each female. The first male will mate and deposit sperm in the female, then a second male will arrive and use its spiny genitalia to scrape out his competitor's sperm, before mating itself.

But that does not mean he has won the competition. Much of the sperm of the first male survives this ordeal, and is carried unwittingly by the second male on its genitalia. If this second male mates again within a short period he can inadvertently inseminate a female with sperm that is not his own; and the first male gets to impregnate a female it has never met. One in eight females are fertilised by proxy in this way.[11]

Virgin male **butterflies** make better lovers than butterflies more experienced in the bedroom arts. Butterflies that copulate for the first time raise a greater number of young than those that have had multiple sexual partners, possibly because they haven't depleted their reserves of sperm.[12]

Male butterflies also use a series of anti-aphrodisiacs to ensure that a female they have mated with does not go on to have sex with another male. Males of the **butterfly** species *Danaus gilippus* cover females in a pheromone-laden dust that contains chemicals that both inhibit her ability to fly, and stick like glue to her antennae, making it more difficult for her to identify other potential male partners. Male *Heliconius erato* **butterflies** also transfer a pheromone to the female, which she then disseminates from special storage organs called 'stink clubs' making her highly distasteful to other males. A similar trick is also used by the **green-veined white butterfly** (*Pieris napi*) and its relative *P. rapae*.

Other males are more obliging. When male **arctiid moths** (*Utetheisa ornatrix*) mate they transfer chemical compounds to the female, which she bequeaths to her offspring, rendering them distasteful and thus protected against predators.[13]

More than 90 per cent of the sperm produced by male **butterflies** and **moths** is non-fertile. Male Lepidoptera produce two types of sperm: normal, fertilising 'eupyrene' sperm, and a much larger number of non-fertile, anucleate 'apyrene' sperm. It is not known exactly what function the non-fertile sperm serve. They might play a role in activating the fertile sperm, or help the active sperm transport themselves through a female's reproductive tract. Alternatively they might provide a source of nutrients either for the fertile sperm, the female or the developing zygotes produced after fertilisation. Most likely is that apyrene sperm play a role in sperm competition, either by displacing or inactivating rival males' sperm, or by remaining in the females' reproductive tract, delaying her ability to mate again.[14]

Females of many spider species, such as **African golden orb web spiders** (*Nephila madagascariensis*), have not one but two genital openings into which males deposit sperm. Not to be outdone, males of the species also have two sexual organs of their own, known as pedipalps, each of which can enter these openings. Unfortunately for the males, these pedipalps often break off while inside the female. The situation isn't good for females either, as this lopped-off sexual organ often prevents them mating later with other better-quality males.[15]

Being eaten by your mate appears to be a good reproductive strategy for **golden orb weaving spiders** of the species *Nephila plumipes*. Males that are eaten after copulating father more offspring than males that are left uneaten. It is likely that this is because females choose to eat only the best quality males.[16]

Some males of the related **orb weaving spider** (*Nephila fenestrata*) try to avoid being eaten by their mates, by approaching a female only once she has caught a fly and is busy with another meal.[17]

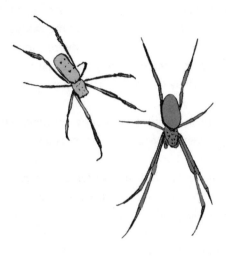

Another spider intentionally chews off one of its genital organs before mating, because they are simply too big and unwieldy. Male **spiders** within the genus *Tidarren* are just 1 per cent of the size of females, yet each of their pedipalp sexual organs makes up 10 per cent of the male spider's body weight. Before mating, males voluntarily remove one of these organs, as the sheer weight and size of their genitals makes running difficult. Males with just one pedipalp run 44 per cent faster and for 63 per cent longer than those with two. They are also capable of running three times as far, a trick that is believed to help the lovelorn male run rings around other males competing for the female's attention.[18]

Female **St Andrew's cross spiders** (*Argiope keyserlingi*) terminate copulation by detaching males from their genital openings, wrapping and then killing them.[19]

Male **bruchid beetles** (*Callosobruchus maculates*) have spines on the tip of their genital organ that are designed to damage a female's reproductive tract during copulation. It is thought that damaging the female this way helps to postpone her mating with a rival male, increasing the chances that the first male will father her young. Females have evolved a counter-adaptation, which involves kicking the male during copulation, a strategy that reduces the damage by shortening the duration of copulation.[20]

Sexual cannibalism is not only one-way traffic. Males of some species will also devour females. For instance, male **amphipods** (*Gammarus pulex and G. duebeni celticus*), a small water-dwelling crustacean, will cannibalise smaller females, especially just after the female has moulted its exoskeleton leaving it at its most vulnerable. A male **paddle crab** (*Ovalipes catharus*) will also eat its female mate during or after mating, again usually just after she has moulted and only has a soft shell. Presumably this happens when the male decides that the probability of producing offspring is quite small, and he has more to gain by eating the convenient meal of the female.[21]

Female **brown trout** (*Salmo trutta*) fake orgasms to encourage males to ejaculate prematurely. By doing so, they dupe their partner into thinking it has successfully mated, before the female fish moves on to find a better male with which to do the real thing.[22]

The **two-spotted goby** (*Gobiusculus flavescens*) has an unusual sex life, as during the early stages of the fishes' short breeding season, males will compete with each other for the attention of females, and try to court prospective mates. But as the breeding season progresses, the situation reverses and females begin to fight over males, and do all the chasing to get their man. The goby is the first vertebrate known to have such a fluid relationship between the sexes.[23]

Small male **bluegill sunfish** (*Lepomis macrochirus*) make up for their weedy size by producing super sperm. The sunfish have two mating strategies. Some males wait until they are seven years old before becoming territorial males that build nests and actively court females. Other smaller males try to mate at a younger age by sneaking up to females and copulating without their partner knowing. But these weedy, sneaky males have a trick up their sleeve that gives them a head start over their larger, more thuggish competitors: they produce physiologically superior sperm that swim faster.[24]

Male **meadow voles** (*Microtus pennsylvanicus*) inseminate females with more sperm when they smell another male competitor nearby. They do not increase the frequency of their ejaculates, but actually produce more sperm within each ejaculate, as a way to increase the chances that their sperm, rather than a rival's, will impregnate the female.[25]

Female **chimps** exhibit large pink swellings on their rear quarters to show they are receptive to mating, the size of which signals their reproductive quality. But these swellings also increase the operating depth of the vagina by up to 50 per cent. In response, male chimps have evolved longer penises to ensure they can deposit sperm closer to the cervix, and have a greater chance of impregnating the female.[26]

Some females prefer losers. Female **Japanese quail** (*Coturnix japonica*) mate with weaker males that habitually lose fights with competitors, rather than the more traditional strategy of mating with the victor. It appears that the females prefer losers in order to avoid being harmed by more dominant aggressive males.[27]

Sometimes the least attractive girls get the most boys. Male **Soay sheep** (Ovis aries), which live feral on the island of St Kilda in Scotland, aggressively compete with each other for females. But instead of mating with as many females as possible the most dominant males preferentially select only the heaviest females which have the greatest chance of successfully giving birth to healthy lambs. That means that the lighter and hence less attractive females are mated by a series of less attractive males, and as a consequence the least appealing females actually mate with more male sheep than their better-looking rivals.[28]

Small can be beautiful. While most females prefer to mate with larger, stronger and therefore fitter males, females of the **bat** species Saccopteryx bilineata, known as **sac-winged bats**, prefer small males. These bats mate on the wing, so it is believed that smaller males can perform better aerial displays and are more able to copulate while flying.[29]

Male **garter snakes** have a tendency to become transvestites, or 'she-males', an odd strategy where young male snakes dress up as members of the opposite sex and attempt to mate with other males. The snakes do this by producing similar lipids on their skin to females. These lipids attract the attentions of other male snakes. All male garter snakes seem to go through this phase as they pass into adulthood, and the reason for becoming a transvestite appears to be twofold; the change appears to switch off the elaborate courtship behaviour males use to woo females, preventing the young snakes from expending too much energy on mating at a time when it is unlikely to be successful. By dressing up like a transvestite, young male snakes may also be gaining an advantage over their older male rivals. If an older snake is tricked into trying to mate with a cross-dresser, it has less chance of fathering young with a female.[30]

Giant cuttlefish (*Sepia apama*) usually live solitary lives lasting just one to two years. But every winter, hundreds of thousands of cuttlefish come together at Spencer Gulf off the coast of South Australia to take part in a huge sexual orgy. Up to 11 males show up for every female, and each uses a variety of tactics to impregnate a partner. Males will guard a female, try to sneak out from behind a hiding place such as a rock to copulate with another's mate, or most strangely, pretend to be another female to avoid being rumbled by the competition.[31]

It is amazing what a moth will do for sex. Both males and female **moths** of the species *Spodoptera littoralis* will normally take evasive action when they hear the echolocating sounds of an approaching hungry bat, a necessary tactic to avoid being eaten. But when a male smells the alluring pheromones of a female, he throws caution to the wind. The prospect of mating essentially causes the male to become deaf to danger and ignore the bat, so intent is he on copulating. The moths only live for two weeks, so it is likely that males weigh up the relative risks of becoming someone's dinner and decide that getting sex while they can and passing on their genes is more important.[32]

A low dose of radiation can affect the sex lives of earthworms. Some **worms**, such as the Japanese species *Enchytraeus japonensis*, usually reproduce asexually, with each individual breaking its body into smaller segments, each of which grows into a new adult worm. But when exposed to low levels of radiation – about 15 times found naturally in the environment – the worms begin to reproduce sexually and lay fertilised eggs that develop into juveniles.[33]

Some birds prostitute themselves for food or shelter. Female **purple-throated hummingbirds** (*Eulampis jugularis*), for instance, will copulate with males in return for being allowed to forage for food within the male's territory, even during the non-breeding season. Female **Adélie penguins** (*Pygoscelis adeliae*) exchange sex for nest material.[34]

And some male birds will keep a mistress, and pay her handsomely to raise their kids. **Great grey shrikes** (*Lanius excubitor*) are a raptor-like passerine bird, and the males and females usually form stable mating pairs. Males like to indulge in a spot of gift-giving, presenting their mates with a prey item such as a rodent, bird, lizard, or large insect immediately before copulation. But some males will have a fling with another bird. What is more, to win over his mistress, the male offers her food that contains four times the energy, and which took much longer to catch, than the food he gives his faithful partner.[35]

One species of mite gets to live longer by having sex. If a female **mite** (*Histiostoma feroniarum*) delays mating by even a few days she will not only have fewer offspring, but is also likely to die more quickly.[36]

Many species of fish attract potential mates by emptying their bladders. Before ovulating, female **goldfish** (*Carassius auratus*), for example, release three steroid hormone-derived pheromones in their urine, which can be smelt by males looking for a suitable partner. The males respond to this cue by increasing the amount of milt – the substance that includes sperm – they produce and by increasing the motility of their sperm. During ovulation, the females release prostaglandins in their urine, which then induces spawning behaviour in the males. Male **tilapia fish** (*Oreochromis mossambicus*) also urinate to advertise their status. Female tilapias are particularly sensitive to the smell of male urine, and males increase their rate of urination as part of their courtship display.[37]

Male **goats** become sexually aroused when they see female goats mount one another.[38]

There are six reliable signs that a female **yak** (*Poephagus grunniens*) is in oestrus. In order of frequency of occurrence they are:

	PER CENT
Following and mounting by male yaks	100
Standing to be mounted	100
Raising of the tail	83
Frequent urination	75
Congestion of the vulvar mucous membrane	75
Swelling of the vulva	67

table source [39]

Unhealthy living

❧

HEDGEHOGS CAN SUFFER from a peculiar disease known as
Wobbly Hedgehog Syndrome. The disease particularly affects
African pygmy hedgehogs (*Atelerix* spp.), which are becoming
increasingly popular pets in North America, as well as in South
America, Europe and some Asian countries, and has also been
reported in wild **European hedgehogs** (*Erinaceus europaeus*). One
in ten pet African pygmy hedgehogs suffer from the syndrome,
which is first characterised by the animals' inability to roll into a
ball and close their hood. The hedgehogs then become increasingly uncoordinated and begin stumbling, tripping or wobbling.
Over several months, the symptoms become progressively more
severe and may include falling consistently to one side, tremors,
seizures, muscle atrophy and self-mutilation. The majority of
affected animals become completely paralysed within 15 months
of the first onset of symptoms. It is not known what causes the
syndrome and there is no treatment or cure.[1]

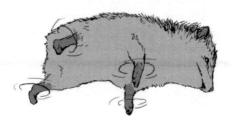

The ongoing rivalry between **cats** and **dogs** has some far-
reaching and unfortunate consequences. In the mid-1970s, a pan-
leucopenia virus used in cat vaccines jumped the species barrier
into dogs, producing a hyper-virulent strain in puppies that

within a few months caused widespread puppy mortality across the world. Payback from the dogs came when a strain of canine distemper, endemic in the pet dogs of Masai tribesman in Tanzania, jumped the species barrier to **hyenas** and then to African **lions,** killing a third of the lions living in the Serengeti National Park within six months in 1994.[2]

Two hundred and fifty-eight genetic diseases of **cats** have so far been described.[3]

Pets, such as **cats, dogs** and **horses**, can pass antibiotic-resistant bacteria to their owners, and vice versa, in particular MRSA strains of *Staphylococcus aureus* bacteria that are resistant to methicillin.[4]

Stickleback fish (*Gasterosteus aculeatus*) infected with parasitic flatworms are more likely to shoal, and form larger shoals, than uninfected fish. Parasitised fish are also more likely to swim at the front of a shoal than non-parasitised individuals.[5]

Honeybees (*Apis melifera*) are electrostatically charged when they are flying and will attract bacterial spores and viruses in the air. These pathogens are then adsorbed onto the honeybee's body.[6]

Sexually transmitted diseases (STDs) are not the preserve of licentious people. Over 200 diseases spread by sexual contact have been documented in animals thus far, plaguing groups as diverse as mammals, birds, reptiles, arachnids, insects, molluscs and nematode worms. STDs can be caused by single-celled protozoa, fungi, nematodes, helminth worms and cancerous cell lines as well as bacteria and viruses.

ANIMAL	DISEASE	TYPE	PATHOGEN
Baboon (*Papio* spp.)	herpes	virus	*Herpesvirus* (Simian agent 8)
Bumblebee (*Anthrophora bombodies*)	nematode infection	nematode	*Huntaphelenchoides* spp.
Cow (*Bos taurus*)	lumpy skin disease	virus	*Capripoxvirus*
Dog (*Canis familiaris*)	canine venereal tumour	dog's own cells	host cell line
Earwig (*Prolabia annulata*)	fungal infection	fungus	*Filariomyces forficulae*
Goose (*Anser* spp.)	goose venereal disease	bacteria	*Mycoplasmosis cloacale*
Horse (*Equus caballus*)	equine herpes	virus	*Herpesvirus*
House mouse (*Musca domestica*)	fungal infection	fungus	*Entomophthora muscae*
Killer whale (*Orcinus orca*)	genital papillomatosis	virus	*Papillomavirus*
Mangabey monkey (*Cerocebus atys*)	Simian Immunodeficiency Virus	virus	SIV
Pig (*Sus domesticus*)	hog cholera	virus	Not known
Python (*Python regius*)	bacterial infection	bacteria	*Aeromonas* spp.
Rabbits, hares and pikas (*Leporidae*)	rabbit syphilis	spirochaete	*Treponema paraluis-cuniculi*

ANIMAL	DISEASE	TYPE	PATHOGEN
Scarab beetle (*Oryctes monoceros*)	nematode infection	nematode	*Oryctonema genitalis*
Sheep (*Ovis aries*)	venereal orf	virus	*Parapoxvirus*
Spider (*Pamphobeteus* spp.)	rickettsial infection	bacteria	rickettsia-like organism
Squid (*Alloteuthis subulata*)	Platyhelminthine worm	worm	*Isancistrum loliginis*

table source[7]

A sexually transmitted disease of **horses** had a major impact on economic, social and military life in Europe in the 19th century. First described by a veterinarian in the district of Trakehnen, Prussia, as a 'malignant disease of the generative organs' of horses, the ailment quickly spread from animal to animal through the Austro-Hungarian Empire, and appeared for the first time in France and Switzerland in 1832. By the latter part of the 19th century, the disease had reached southern Russia and northern Africa. Known only to infect stallions and mares, and never geldings or foals, the disease decimated horse populations, killing 600 horses owned by the Rigas Arabs alone in 1847. It was not until 1894 that the causative agent, *Trypanosoma equiperdum*, was identified, and over the coming decades quarantine and slaughter policies eliminated the disease.[8]

Between 5 and 75 per cent of female **koalas** (*Phascolarctos cinereus*) may be infected at any one time by chlamydia, a sexually transmitted disease caused by the bacterium *Chlamydia* *psittaci*. The koala pathogen is closely related to *Chlamydia trachomatis*, which also causes chlamydia in humans.[9]

Nine-banded armadillos (*Dasypus novemcinctus*) in Texas and Louisiana can contract human leprosy, while armadillos in Florida are immune to the disease.[10]

There is one species of lobster that will shun its neighbour if it is carrying a potentially lethal disease. **Caribbean spiny lobsters** (*Panulirus argus*) are usually gregarious and like to share underwater dens with one another. But normal, healthy individuals will avoid sharing a home with other lobsters that carry a lethal virus that can be passed by direct contact. It is not clear how a lobster knows that its neighbour may be infected. However, it will begin to shun its diseased compatriot within four to eight weeks of it contracting the virus, just before the diseased individual becomes contagious.[11]

Drink and drugs

❧

Monkeys LIKE A drink or two. Given the choice of whether to have an alcoholic beverage, or something teetotal, around one in 20 **vervet monkeys** (*Cercopithecus aethiops*) become instant binge drinkers, gulping down so much booze that they eventually pass out. Around one in seven are heavy drinkers who like their spirits neat, while most are moderate drinkers who prefer to wash down their alcohol with a little fruit juice. Just one in seven monkeys decide not to have an alcoholic drink at all.[1]

Monkeys also use their taste for alcohol to work out whether fruit will be good to eat. Fruits contain low levels of ethanol, and, by smelling the juicy treat, a primate can tell how much ethanol is inside, which in turn is a good guide to whether the fruit is ripe.[2]

Chimpanzees (*Pan troglodytes*) self-medicate. The apes will fold up and eat rough bristly leaves, not for any nutritional value, but to clear out parasites such as intestinal worms in their digestive tract. The leaves and many of the parasites are then excreted.[3]

Sir William Osler, supposedly the most famous physician at the turn of the 20th century, and one of the great icons of modern medicine, said, 'The desire to take medicine is perhaps the greatest feature which distinguishes man from other animals.' Not so. When given the opportunity, **sheep** (*Ovis aries*) will preferentially eat medicinal compounds in order to self-treat stomach ailments. Sheep fed grain appear to prefer foods and solutions that contain substances such as sodium bicarbonate that attenuate acidosis. In the first direct evidence of its kind in the animal kingdom, it has been shown that lambs can also learn to associate multiple illnesses with a range of specific medicines. What is more, when they fall ill they will ingest the specific medicine required to make them feel better. Lambs conditioned to eat grain, tannins, and oxalic acid, each of which leads to digestive problems, will learn to eat sodium bentonite, polyethylene glycol or dicalcium phosphate respectively, choosing the specific medicine that will alleviate their condition. Once they have taken their medicine, their health improves as the animals recover their appetites.[4]

Cows in the USA and Mexico routinely get smashed on crazy weed, a poisonous plant that grows in their fields. First they become lonesome, then they start walking funny, until eventually they start bumping into things. Some cows that are as high as a kite even take giant leaps over the smallest obstacles such as a stick. Eventually they freak out so much that they become impossible to control and will charge anyone who enters their field.

Crazy weed, or locoweed, as it is also known, is actually a collection of plant species within the genus *Astragalus* and *Oxytropis*, and the cows love it. Despite it being toxic the animals munch away and will even encourage other members of the herd to give it a try. The plant contains an alkaloid called swainsonine that stops the cow's body breaking down glycoproteins. Eventually these proteins build up and damage each cow's nervous system, making the animal appear crazy.[5]

One-third of **cats** do not react at all to catnip, the mint-like plant *Nepeta cataria* that usually makes Tiddles go crazy. Lions, tigers and lynx as well as domestic cats all love the plant, but one in three cats lack a gene required to sense the active chemical within catnip, a terpene called nepetalactone.

Cats in Japan love a different drug altogether. The matatabi plant contains a similar compound to nepetalactone, but its effects on Japanese cats are somewhat different. Whereas catnip makes cats chew, shake, roll around and drool, the matatabi plant makes Japanese cats lie on their backs and stick their paws in the air.[6]

There is anecdotal evidence that birds sometimes get sloshed on fermented berries, and do a spot of drunk flying. The result can be fatal, with particularly **robins** and **cedar waxwings** reportedly weaving about in the air before crashing into windows or trees, or falling to their deaths from their perch.[7]

Fruit flies fall asleep when they get drunk on a nip of ethanol. But there is one mutant type of **fruit fly** (*Drosophila* spp.) that does not feel the ill effects of a bout of boozing. These flies, lacking neurons in their brain that respond to a signalling factor called neuropeptide F, are relatively immune to the effects of ethanol, and remain wide awake after a drink of 42 per cent ethanol, a concentration that puts most flies to sleep.[8]

Hungry **rats**, adolescent rats, or rats previously exposed to drugs are more likely to become addicted to nicotine.[9]

Hunters in the Amazon region of Peru often give their hunting **dogs** hallucinogenic drugs in an effort to enhance their hunting capabilities.[10]

Talented creatures

⚬

DOLPHINS (*Tursiops truncatus*) can point. When they discover a novel object, they have been known to communicate its whereabouts to human scuba divers by aligning their bodies with the object, then moving their heads from the diver to the object and back again several times.[1]

Chimpanzees (*Pan troglodytes*) can understand the different values of Arabic numerals, and then remember them for at least three years.[2]

Lions appear to have a rudimentary ability to count. When a pride of lions hears the roar of a single approaching **lion** (*Panthera leo*) two or three females, rather than a lone greeter, will always go out to meet the stranger. But if two approaching lions can be heard, then the resident females up the ante, sending out four of their own, and so on. In this way, the big cats seem capable of keeping track of how many companions and strangers are around.[3]

New Caledonian crows (*Corvus moneduloides*) are advanced toolsmiths. Not only are they able to select the appropriate length of stick for the task at hand, such as levering grubs out of holes in wood, they are also able to select the stick with the appropriate diameter to fit into the hole. The birds can even manufacture a tool the right length and width tool from twigs pulled from trees, an ability only previously seen in some finches and primates.[4]

A **raven** (*Corvus corax*) will intentionally deceive another raven and lead it away from a source of food so that it can return and eat the meal itself.[5]

Burrowing owls (*Athene cunicularia*) use parcels of poo to seduce their prey and lure them closer. The owls collect up the droppings of larger mammals and then transport them back to their burrow entrance, where they arrange the faeces. The poo acts as bait for dung beetles, the owls' main source of food, drawing them close to the burrow where they are picked off and eaten by the owls.[6]

Green herons (*Butorides virescens*) have been recorded placing floating objects such as bread, feathers and insects onto the water surface, using them as bait with which to lure fish, which the heron then tries to eat.[7]

Pig-tailed macaques (*Macaca nemestrina*) will notice if a human is imitating their monkey behaviour, but are not clever enough to know how or why someone is copying them.[8]

When a **chimpanzee** is confronted by a problem that it cannot solve by itself it will actively recruit another chimp to act as a collaborative partner. What is more, a chimp is able to identify which other chimp will best help it solve the problem and choose that helper over others less able to help it solve the task.[9]

The **sea otter** (*Enhydra lutris*) is the only mammal apart from primates known to use tools naturally. The otters lie on their backs and use pebbles or small rocks held with their paws to smash open shellfish such as abalones and mussels.[10]

Rats can learn the difference between Dutch and Japanese. If the animals are rewarded each time they hear one or other of the languages, they soon learn to distinguish the two. They can spot the difference even when the languages are spoken backwards, suggesting the animals recognise the rhythmic patterns of each.[11]

They can also be trained to work as sniffer dogs. The rodents are so good that after just a couple of weeks' training they have a 90 per cent success rate in sniffing out illicit hoards of cocaine.[12]

In a certain way, hyenas are just as clever as people or monkeys. Primates were the only group of animals thought able to recognise third-party relationships. For instance, a primate will work out whether one animal is dominant over another and use that information when making allies or reconciling after conflicts. It turns out that **spotted hyenas** (*Crocuta crocuta*) are capable of just the same. Hyenas are carnivores that live in gregarious communities according to strict dominance hierarchies. Spotted hyenas will always support the dominant animal in a fight between pack members, even if both the fighting animals rank above or below the observer. Hyenas are also more likely to attack relatives of their opponents once a fight has ended, showing that they are capable of quite Machiavellian manoeuvrings.[13]

The **rufous hummingbird** (*Selasphorus rufus*) is the only animal known that can remember both the locations of food sources in the wild and when they last visited them. The ability to remember not only where nectar-laden flowers are, and when the hummingbird last took a meal from them, means the bird is able to avoid returning until the flowers have refilled with nectar.[14]

Honeybees (*Apis mellifera*) can recognise individual human faces.[15]

Young **calves** less than three weeks old are able to distinguish between different people, an ability heightened if the people regularly wear different colour clothes.[16]

Monkeys will pay to look at images of other sexually attractive or powerful primates. When tested, male **rhesus macaques** (*Macaca mullatta*) will give up a drink of appetising cherry juice in order to view a picture of a face of a socially dominant macaque, or of a female's hindquarters. However, the monkeys have to be paid in kind to look at a picture of a subordinate, as they only take a peek if they are bribed with a larger than normal drink.[17]

Some species of monkey intentionally urinate onto their hands. For instance the **moustached tamarind** (*Saguinus mystax*) occasionally washes in its own urine as well as splashing it onto its palms. It is not clear why they do this, but it may play a role in regulating the animals' temperature or help them conserve water, act as a cleaning aid, help improve their grip, or act as a social signal to others.[18]

Ants are specialised gardeners, helping plants of the tropical forest take seed and grow. **Worker ants** of species such as *Pheidole plagiaria* descend on discarded seeds on the tropical forest floor, eating all the surrounding fleshy pulp. They then leave the hard seeds but not before coating them with an anti-fungal agent, which helps them germinate faster and more successfully than seeds left alone.[19]

Domestic pigs wallow to cool down, because they do not have many sweat glands in their skin. However, for the **wild boar** (*Sus scrofa*) wallowing may serve a different function, being a way for the boar to advertise its sexual prowess. Wild males that frequently wallow and have a layer of mud on their skin are larger and more sexually active than those that do not.[20]

Unlike most of its relatives, the **Namib desert golden mole** (*Eremitalpa granti namibensis*) does not dig tunnels underground. Instead it is known as a sand swimmer, because the golden mole burrows through loose sand which collapses behind it rather than producing tunnels in cohesive soil. While underground, the mole is completely surrounded by sand.[21]

Australian fiddler crabs (*Uca mjoebergi*) will gang together to see off a larger rival. The crabs do this even with unrelated neighbours, the first example of an animal strategically helping a neighbour to defend its territory against an intruder. This neighbourhood watch scheme makes sense because it takes less effort for a crab to join forces with another than it does to take on the intruder itself and renegotiate territorial boundaries on tidal mudflats with the newcomer.[22]

Talking ...

Bottlenose dolphins (*Tursiops truncatus*) can mimic the whistle sounds of other dolphins they hear underwater, being one of the few animals known that can imitate the calls of their compatriots.[1]

Baby bottlenose dolphins only learn to 'speak', or develop their own unique signature whistles, when they are between one and a half and two and a half months old.[2]

Humpback whales (*Megaptera novaeangliae*) use their own syntax and grammar in the songs they sing. Humpbacks sing complex patterns of moans, cries and chirps and produce song elements, or phrases, embedded in larger, recurring themes. Such a hierarchical structure in language is also used by another animal: humans, who use sentences which contain clauses, which in turn contain words.[3]

Richardson's ground squirrels (*Spermophilus richardsonii*) whisper to each other to warn that a predator might be approaching. The squirrels produce audible alarm calls at a frequency of about 8 kilohertz, but they also emit 'whisper' calls, which contain pure ultrasonic frequencies of around 50 kilohertz, one of the few examples known of an animal using ultrasound as a warning signal. Because the sound of these whispers attenuates quickly and tends to be highly directional it may be that the squirrels only whisper to nearby relatives, sending them a discreet warning that cannot be heard either by a predator or unrelated squirrels further away.[4]

Mice sing to one another, and they do it in ultrasound. Male **laboratory mice**, when they encounter female mice or their pheromones, emit ultrasonic vocalisations with frequencies ranging from 30 to 110 kilohertz, while pups produce 'isolation calls' when cold or removed from the nest. These vocalisations of male mice have all the characteristics of a song, as sung by birds for example. The ultra-high-pitched squeals of a male mouse contain discrete, rich elements including several types of syllable organised into phrases and motifs. A syllable is a unit of sound separated by silence from other sound units, and it may consist of one or more notes. A phrase is a sequence of syllables uttered in close succession, whereas a motif is a sequence of several syllables of a particular type, where the entire sequence is observed repeatedly in the animal's vocalisation. A single mouse can utter 750 syllables within a period of just three and a half minutes and individual male mice sing their own particular serenades to females preferring certain syllables over others and structuring their songs differently over time. Mice are the only other mammals apart from whales, bats and humans to sing to one another.[5]

Chimpanzees (*Pan troglodytes*) have between 20 and 30 ways of vocally or facially expressing themselves that can be variously categorised as barks, screams, grunts, pants, hoots and pouts. Specific vocal expressions include yelps, squeaks, shrill barks, pant-grunts, pant-hoots, pout-moans and the stretch pout-whimper. Specific facial expressions include the relaxed lip face, bulging lip face, relaxed open-mouth face, or play face.[6]

Elephants have over 30 different ways to talk to each other. So far, **African** and **Indian elephants** (*Loxodonta africana* and *Elephas maximus*) have been recorded making 31 types of call, most of which are low-frequency infrasound noises that are imperceptible to the human ear. For instance, different calls are used to assemble a group of adults and calves, or to call to herd members long distances away.[7]

Horses (*Equus callabus*) make six distinct recognisable sounds: the scream, squeal, nicker, whinny, snort and blow.[8]

Dogs (*Canis familiaris*) consistently bark in at least three different ways and each type of bark is used in a specific context. For instance, when a dog is disturbed it will bark for longer and more intensely and with a harsher, lower frequency than when it is playing or alone. Then it will tend to produce a more tonal bark at a higher pitch. The same is true across different breeds and sometimes the intervals between the barks of a disturbed dog are so short that they fuse together into what scientists call a super-bark. Researchers also suspect that dogs may bark differently depending on whether they are playing with a person they know or a stranger. Individual dogs can also be identified by their barks and there is some evidence that **wolves** (*Canis lupus*) can be individually recognised by their howls.[9]

Domestic **cats** purr at a frequency of 26 hertz, a cheetah at 17.5 hertz and a puma at 16.8 hertz. No one is sure why cats purr.[10]

Cat species that purr

Asian golden cat (*Catopuma temminckii*)
Black-footed cat (*Felis nigripes*)
Bobcat (*Lynx rufus*)
Cheetah (*Acinonyx jubatus*)
Eurasian lynx (*Lynx lynx*)
Indian desert cat (*Felis silvestris ornate*)
Jaguarundi (*Herpailurus yaguarondi*)
Leopard cat (*Prionailurus bengalensis*)
Marbled cat (*Pardofelis marmorata*)
Margay (*Leopardus wiedii*)
Ocelot (*Leopardus pardalis*)
Puma (*Puma concolor*)
Serval (*Leptailurus serval*)
Tiger cat (*Leopardus tigrinus*)
Wild cat/domestic cat (*Felis silvestris*)

Species for which there is no evidence they purr

Clouded leopard (*Neofelis nebulosa*)
Jaguar (*Panthera onca*)
Leopard (*Panthera pardus*)
Lion (*Panthera leo*)
Snow leopard (*Uncia uncial*)
Tiger (*Panthera tigris*) [11]

Songbirds stutter. Just as some people struggle to complete a whole word, around 7 per cent of **zebra finches** (*Taeniopygia guttata*) repeat the same part of their song over and over again before breaking out and continuing their tune. In both people and birds, a stutter is thought to be caused by disturbances in the motor control regions of the brain.[12]

A few species of bird, such as the **bay wren** (*Thryothorus nigricapillus*), will sing duets, where a male and female coordinate their songs as a way of demonstrating their commitment to one another.[13]

Chaffinches (*Fringilla coelebs*) living in noisy places, such as near waterfalls or bustling rivers, sing the same song over and over again to make sure they are heard over the din. So far, they are the only bird known to repeat parts of their song to avoid being drowned out by surrounding noise.[14]

Black-capped chickadees (*Poecile atricapilla*) produce specific alarm calls that tell other birds the size of the predator approaching. The birds emit more 'chick-a-dee' alarm calls when attacked by a small predator such as a **pygmy owl** (*Glaucidium californicum*) than they do when attacked by a larger predator such as a larger great-horned owl (*Bubo virginianus*).[15]

Ringed doves make their familiar cooing sound by using super-fast muscles that surround their syrinx, a vocal organ that is unique to birds. The **doves** (*Streptopelia risoria*) use their syrinx to generate a relatively simple cooing song that also contains a repeating trill that occurs at high frequencies of around 30 hertz. To make the trilling sound, the doves pass air through their respiratory tract exciting membranes in the syrinx and causing them to vibrate. The frequency of this vibration is controlled by tension in the membranes, which in turn are controlled by two pairs of syringeal muscles. However, normal vertebrate muscles cannot contract fast enough to produce the short sound elements within each trill, which can last for less than 9 milliseconds. The doves' syringeal muscles, however, are similar to those that rattlesnakes use to vibrate their tails, and are among the fastest-acting muscles in the animal kingdom.[16]

Echolocation helps bats both to navigate and home in on their insect prey. But it may also be a way for the creatures to communicate. **Big brown bats** (*Eptesicus fuscus*) can recognise whether the echolocation calls of a neighbour are being produced by either a male or a female. Females tend to respond far more strongly to the calls of other females, while ignoring the calls of males. No one knows why, but it may be that such communication is important for female big brown bats, which form social hierarchies when raising their young in roosts.[17]

and listening ...

It can take a **goldfish** (*Carassius auratus*) a month to recover its hearing after it has been exposed to a loud noise.[1]

The **common** or **Dover sole** (*Solea sole*) helps orientate itself by listening to the noise produced by the wind above the waves.[2]

Fish have a wide range of hearing abilities. Two-thirds of freshwater fish, including species of carp, are hearing specialists. They can hear sounds over a broad frequency range of up to several kilohertz and at much lower sound intensities than the remaining third of species known as hearing generalists. They include salmon and perch, which can only detect low-frequency sounds at less than 1 kilohertz at relatively high sound intensities.[3]

Java sparrows (*Padda oryzivora*) appear to prefer the music of some composers compared to others. Sparrows will listen longer to music composed by Bach than by Schoenberg, and prefer Vivaldi over Carter.[4]

Mammals evolved their specialised way of hearing twice. Unlike birds and reptiles, mammals have three bones in the middle ear, known as the hammer, anvil and stirrup, which process sound. This complex way of hearing evolved once in placental and marsupial mammals, and again in a third group of mammals known as monotremes, which include the **platypus** (*Ornithorhynchus anatinus*).[5]

Echolocating bats, mice, ground squirrels, whales and dolphins are not the only animals capable of hearing ultrasound. Fish can, too, as both **blueback herring** (*Alosa aestivalis*) and **American shad** (*Alosa sapidissima*) respond to sounds up to 180 kilohertz.[6]

There is also one species of frog that communicates using ultrasound, the only amphibian known to do so. The concave-eared **torrent frog** (*Amolops tormotus*) emits sounds at a frequency of greater than 20 hertz, which can be heard by other frogs over the noise produced by nearby streams.[7]

Eating ...

∾

THERE ARE MOTHS that drink the tears of elephants. Others drink from the tears of buffalo, while yet more feed from the wet eyes of horses, deer, tapirs and pigs. And each moth's life is dependent on the salty secretions wept by their hosts. This remarkable behaviour astounded entomologists when first discovered. But now we know that there are many moth species that play the crying game, and have formed an intimate relationship with their weeping hosts.

Strangely, **moths** have never been seen to feed from the tears of carnivores such as dogs or large cats, from birds, or from other groups of vertebrates such as marsupials. They prefer the tears of large hoofed mammals, elephants, and on occasion people, whose eyes will often be visited at night while asleep. It is not unknown for a slumbering person to wake a number of times during the night to brush aside a gentle visitor trying to stealing a meal from the corner of their eye.

Why do the **moths** do it? Tears contain both salt and water, two precious commodities for any animal. Tears also contain trace levels of proteins. And while moths usually like a sugary meal, some are able to break down protein in their guts, making tears an unlikely but potentially very nutritious source of food.

Some species of **moth** engage in an almost romantic kissing of the eyes. *Mabra elephantophila*, for example, which drinks the tears of elephants, is among the smallest of such moths. A shy, delicate creature, its tiny size allows it to steal a tear without the elephants seemingly noticing.

Tarsolepsis remicauda is a much larger **moth**, with a wingspan of some 80 millimetres. Its silvery wings, and long red bushy abdomen make it hard to ignore, and even weeping animals notice its flamboyance and large size, and attempt to brush the moth aside.

Others are a little more sinister, however. The highly specialised *Lobocraspis griseifusa* does not wait for an animal's eyes to moisten. When it has landed, it sweeps its proboscis across the eye of its unfortunate host, irritating the eyeball, encouraging it to produce tears. It can even insert its proboscis between the eyelids, ensuring it can feed even while its host is sleeping. Whereas a **moth** of the genus *Poncetia* goes to the opposite extreme. Its proboscis is so short it must cling to the eyeball itself to drink. But it must be careful. If its weeping host blinks, the moth is often crushed to death.[1]

Cats (*Felis silvestris*) lack a sweet tooth and are unable to taste sweet foods. Over the course of evolution cats lost the genetic material that codes for a vital protein that makes up the sweet tasting receptors possessed by other mammals. **Cheetahs** (*Acinonyx jubatus*), **tigers** (*Panthera tigris*) and their more distant relation the **hyena** (*Hyaena hyaena*) similarly lack a sweet tooth. Cats are also deficient in an enzyme that digests sucrose. If fed water containing sucrose a cat will become violently ill, but because it cannot taste sugar it will not learn to avoid sweet drinks and will repeatedly fall ill as a result.[2]

Several species of primate, including **mountain gorillas** (*Gorilla beringei beringei*) living in Uganda, eat or lick dead or decaying wood, a puzzling behaviour considering that wood contains little if any obvious nutritional value. Wood is low in protein and sugar, and high in lignin compared to other foods.

The reason they eat wood is because it is a vital source of sodium. Whereas people are generally advised to eat a low sodium diet, many primates require high levels of it to remain healthy, and gorillas obtain 95 per cent of their dietary sodium from wood that constitutes less than 4 per cent of their total food intake.

Gorillas have such a fondness for this food supplement that they will often suck on wood for several minutes before spitting it out, or will lick the bases of tree stumps and the insides of decaying logs. Some individuals frequently break off pieces of wood to carry and eat later while others suffer from bleeding gums having chewed wood for long periods of time.[3]

Chimpanzees (*Pan troglodytes*) taste bitter compounds differently from how people do. The ability to taste bitter compounds helps animals regulate how many toxic plants and substances they eat, ensuring they can eat nutritious but toxic foods without succumbing to poisoning or death. However, the genetic mutations that control the taste of bitter substances in chimps and humans are different and evolved separately.[4]

Male **domestic chickens** (*Gallus gallus*) will pretend to offer dinner to a female chicken in order to lure her to his abode. Males regularly produce specific calls that indicate they have found food, even when no food is available, and the deception lures females closer, giving the male a chance to mate with a new unsuspecting partner.[5]

Great tits (*Parus major*) grow fat when there is little chance they will be eaten by predatory birds of prey, and lose weight when the predators show up again. When populations of the **Eurasian sparrowhawk** (*Accipiter nisus*) decline, for instance, great tits pile on the pounds. When sparrowhawk numbers increase, the tits slim down, probably to ensure they are more manoeuvrable in the air if they ever come under attack and have to fly for their lives.[6]

Dracula ants (*Adetomyrma venatrix*) suck the blood of their young. Queen Dracula ants, which belong to the rare genus *Adetomyrma* and live in Madagascar, cut holes into their own larvae and feed upon the hemolymph, or insect blood, that oozes out, thus earning them their nickname.[7]

Vampire bats are unique in more ways than one. As well as feeding exclusively on the blood of live jungle animals, **common vampire bats** (*Desmodus rotundus*) have not learnt to avoid badly tasting foods. All other bats, and all mammals, learn to avoid sour, toxic foods as a way of preventing being poisoned. The vampire bat however, does not, perhaps because it never encounters such toxins in its highly specialised diet.[8]

Male **king penguins** (*Aptenodytes patagonicus*) can do something no other higher vertebrate can – store undigested food in their stomachs for up to three weeks. The birds are thought to produce an antibacterial chemical that kills off nasty bugs in their digestive systems, allowing the swallowed food to stay fresh until it can be regurgitated for its chick.[9]

Cleaner fish (*Labroides dimidiatus*) are supposed to clean their hosts, darting in and out of their mouths and gill slits to remove small parasitic arthropod pests. But given the chance, the cleaner fish actually prefer to eat the mucus within the mouths of the

parrot fish (*Chlorurus sordidus*) they are supposedly cleaning, rather than doing their job of picking off parasites.[10]

The odd-looking **star-nosed mole** (*Condyhru cristutu*) is the fastest-eating mammal in the world, capable of wolfing down a snack of worms in just 227 milliseconds. The mole's ability stems from the 22 pink fleshy tentacles that adorn its face, each highly sensitive to touch. The tentacles allow the mole to recognise what is and is not food within its underground tunnels incredibly quickly. The animal also has a brain well adapted to speed eating, containing three areas within the cortex that are devoted to touch, compared to two for most other species of mole. A hungry star-nosed mole may be reacting close to the limits of how fast its nervous system can process information.[11]

Tyrannosaurus rex, one of the largest predatory **dinosaurs** known, certainly knew how to eat. A growing *T. rex* was able to put on up to 2.1 kilograms a day, a growth rate unparalleled among its closest relatives.[12]

There is one species of snake that doesn't eat its prey whole. **Gerard's water snake** (*Gerarda prevostiana*), a species that lives in south Asia, eats freshly moulted crabs. Because each crab is much larger than the snake, the snake coils its body around its hapless victim and then bites into the soft shell until it has torn a chunk off, the only one of around 2,700 snake species to use such a technique.[13]

A **Burmese python** (*Python molurus*) will increase its metabolic rate 40-fold when digesting a meal. Within two days of feeding, the snake's kidneys, liver, pancreas, lungs, heart and stomach all substantially increase in mass. In particular, the python's heart ventricles grow by up to 40 per cent, allowing oxygen consumption to increase sevenfold. Post digestion, the mass of all these organs returns to normal as the animal fasts while waiting for its next meal.[14]

Lactating **mice** (*Mus musculus*) eat more than twice as much as virgin mice.[15]

European eels (*Anguilla anguilla*), or **silver eels** as they are also known, will not eat at all during their long migration to the Sargasso Sea.[16]

Spiders are incapable of eating solid food, and they do not ingest their prey whole. In this respect they are different from most venomous creatures such as snakes or cone snails. Spiders have to first liquefy their prey. They then siphon off vital juices from the dead victim into their mouth using a specialised sucking stomach.[17]

No wonder Bambi grew up to be king of the forest. Female **red deer** (*Cervus elaphus*) produce more milk, with higher protein content, for male calves compared to female offspring.[18]

A female **hippopotamus** (*Hippopotamus amphibious*) can eat more food relative to its body size than any another ruminant. Its appetite is so big that its stomach contents can make up one quarter of its total body weight.[19]

Donkeys and **ponies** that are providing milk for their young take half as many bites of grass again when grazing as those that are not lactating.[20]

A diet high in unsaturated fatty acids allows **birds** to exercise more strenuously than a diet rich in saturated fatty acids. It is suspected that song birds put on body reserves of unsaturated fat before migrating.[21]

Grazing **sheep** and **cattle** prefer a diet that is 70 per cent grass and 30 per cent legumes. However, their tastes vary during the day, with sheep and cattle preferring to eat relatively less grass and more legumes in the morning and then progressively higher proportions of grass as the day goes on. Lactating animals also generally show a higher preference for legumes than non-lactating animals.[22]

The body mass and length of jaw bone of a **mammal** determine how quickly it chews its food. Small mammals chew more rapidly than larger mammals, whereas mammals with longer jaws chew more slowly. Specifically, chewing frequency is proportional to body mass to the power of -0.128.[23]

Chewing frequency in declining order from fastest chewers to slowest:

Guinea pig (*Cavia porcellus*)

Rat (*Rattus norvegicus*)

Cat (*Felis domesticus*)

Dog (*Canis familiaris*)

European wild hog (*Sus scrofa*)

Rhesus monkey (*Macca mulatta*)

Giant Indian fruit bat (*Pteropus giganteus*)

Cow (*Bos taurus*)

Horse (*Equus callabus*)

African elephant (*Loxodonta africana*) [24]

African elephants (*Loxodonta africana*) use between one and four teeth to chew their food.[25]

African lions' (*Panthera leo*) favourite delicacies are in order of preference: gemsbock, buffalo, blue wildebeest, giraffe and zebra, which are all preyed upon at a greater rate than would be expected considering the densities of their populations. Lions' least favourite meals are, in order of the most distasteful first: baboon, elephant, impala, reedbuck, Grant's gazelle, ostrich, bushbuck and Thomson's gazelle, which are eaten infrequently relative to their population densities.[26]

An **orang-utan** (*Pongo pygmaeus abelii*) will chew a single piece of fruit for up to 20 minutes before spitting out the seeds.[27]

Bison (*Bison bison*) meat has been frozen by human hunters for the past 8,000 years within the cold lava tube caves of southern

Idaho. A single bison yields an average of 354 kilograms of meat, far too much to be eaten in one sitting.[28]

Male **pipefish** (*Syngnathus typhle*) show more interest in food than they do in prospective mates, whereas females are the opposite, being more interested in prospective partners than food.[29]

Giant clams (*Hippopus* spp. *and Tridacna* spp.) grow significantly faster and have a higher chance of survival if grown in effluent rather than in normal seawater.[30]

A species of crab that lives around underwater hydrothermal vents feeds on the equivalent of sea snow. The **crab** (*Xenograpsus testudinatus*) lives in waters off the coast of Taiwan that are normally poor in nutrients due to the high levels of sulphur compounds emitted by the vents. But every now and then, the underwater currents fade away and the vent plumes rise up vertically instantly killing any animals in their path. When this happens, thousands of crabs quickly emerge from their hiding places to feed on millions of dead zooplankton that rain down in a whitish blizzard that looks like falling snow, a behaviour not seen in any other organism living around hydrothermal vents.[31]

All known **reptiles** and **amphibians** that live underground are carnivorous, eating a diet rich in earth-living animals such as termites, ants and worms.[32]

There is no species of bat that feeds exclusively on other bat species. A very small number will take other bats as part of a more varied diet, including the **spectral bat** (*Vampyrum spectrum*), which is the largest bat in the New World. With a wing span of 1 metre, and long canine teeth, the spectral bat will occasionally take other bats as well as feeding on birds and small rodents. The **ghost bat** (*Macroderma gigas*), the only carnivorous bat in Australia, will also feed on bats given the opportunity, dropping on its prey from above, enveloping it in its wings and killing with bites to the head and neck. Other species that eat bats as part of a varied diet of rodents, lizards and frogs are the **big-eared woolly bat** (*Chrotopterus auritus*) and the **heart-nosed bat** (*Cardioderma cor*), the **greater false vampire bat** (*Megaderma lyra*) and the **large slit-faced bat** (*Nycteris grandis*).[33]

Laboratory mice fed a high fat diet become more and more sleepy the fatter they get.[34]

... and excreting

Penguins have an extraordinary ability to shoot poo. A study of the viscosity of penguin faeces, and how far and over what trajectories the birds expel them, shows that penguins exert a pressure of up to 60 kilopascals, a force four times that used by people having a poo. The bird expels its faeces with such force to avoid soiling its nest or feathers.[1]

Penguin guano threatens to destroy the oldest settlement on Antarctica. **Adélie penguins** (*Pygoscelis adeliae*) have left a metre-high wall of excrement around two wooden huts erected by Norwegian explorer Carsten Borchgrevink in 1899 on a beach at Cape Adare. The guano helps feed fungi that are eating the buildings' walls and ceilings.[2]

Oysters produce fake poo. Like many other filter-feeding bivalves, oysters produce pseudo-faeces, consisting of marine sediments that are trapped and filtered, but never pass through the animals' guts.[3]

Cave-dwelling **bats** excrete so much guano that it can form a layer 10 centimetres deep with one year.[4]

Ruminant livestock such as **cows** or **sheep** fart or belch out so much methane that they account for at least 54 per cent of all New Zealand's greenhouse gas emissions. Australian ruminants account for 12 per cent of their country's emissions.[5]

Sheep waste between 2 and 15 per cent of their total energy intake farting or belching out methane.[6]

Mosquitoes don't just bite; they also treat us to the ultimate indignity of urinating on us after they have ingested their meal of blood. Female **mosquitoes** (*Anopheles gambiae*) intake so much blood that they swell up to twice their normal size, which makes taking off into the air and flying difficult. To get around this problem a mosquito pumps out all the unwanted water and salts from the blood it has just eaten by excreting vast amounts of sodium-rich urine.[7]

Species that regularly eat their own faeces[1]

∾

Rabbits and hares
Domestic rabbit (*Oryctolagus cuniculusi*)
Japanese hare (*Lepus brachyurus*)
Mountain hare (*Lepus timidus*)

Pikas
Afghan pika (*Ochotona rufescens*)
Northern pika (*Ochotona hyperborea*)

Primates
Sportive lemur (*Lepilemur mustelinus*)

Marsupials
Common ringtail possum (*Pseudocheirus peregrinus*)
Coppery ringtail possum (*Pseudochirops cupreus*)
Greater glider (*Petauroides volans*)
Koala (*Phascolarctos cinereus*)

Rodents
Chinchilla (*Chinchilla lanigera*)
European beaver (*Castor fiber*)
Guinea pig (*Cavia porcellus*)
Laboratory rat (*Rattus norvegicus*)
Laboratory mouse (*Mus domesticus*)
Meadow vole (*Microtus pennsylvanicus*)
Norway lemming (*Lemmus lemmus*)

Breathe easy

SEALS HOLD THEIR breath while sleeping on the surface of the water.[1]

Fruit flies (*Drosophila* spp.) stick out their proboscis while flying. This strange behaviour helps the fly breathe. The proboscis was thought to be used just for eating, but the flies stick out their equivalent of a tongue while on the wing and then move it around. This movement allows the proboscis to act as a pump, helping to draw air into the fly's body.[2]

For insects such as **butterflies** and **grasshoppers**, breathing oxygen can be fatal. While they need some of the gas to stay alive, breathing too much can kill them. Animals such as fish and mammals transport oxygen around their bodies using proteins such as haemoglobin, which allows them to limit how much of the gas reaches their muscles.

However, insects cannot do this as they breathe through a system of branching tubes that delivers oxygen straight to their tissues. If too much of the gas gets in, it becomes toxic. To prevent this happening, many insects close the valve-like openings to these tubes, called spiracles, for over 15 minutes at a time. In effect, they stop breathing for long periods to avoid being posioned by the very air that keeps them alive.[3]

A **dormouse** (*Glis glis*) entering into a state of hibernation will take up to 260 breaths per minute.[4]

Under pressure

∾

MICE HOUSED IN red cages are far more anxious than those kept in white cages. Mice prefer white cages, then black or green, with red being the least liked. Anxious mice spend longer in the corners of their cages, are more flighty and difficult to catch. This is despite the fact that we still do not know for sure whether mice have colour vision. Mice also prefer opaque cages to transparent ones, polypropylene to wire cages, and long cages over square ones.[1]

Laboratory **rats** exposed to severe stress develop a similar syndrome to depression and post-traumatic stress disorder.[2] Also, rats allowed to exercise on a running wheel are less likely to become depressed when stressed than rats that are not able to exercise.[3]

European breeds of **sheep** become less stressed when transported than other ruminants.[4]

Stressed **ring-tailed lemurs** (*Lemur catta*) are six times more likely to die in the wild in Madagascar than non-stressed lemurs.[5]

When frightened or startled, **pygmy sperm whales** (*Kogia* spp.), a much smaller relative of the sperm whale, discharge a reddish-brown fluid from their intestines to confuse any approaching predators such as sharks. This fluid muddies the surrounding water and is thought to be used in the same way that an octopus or squid shoot out dark ink.[6]

You can predict whether a **bull** will react to a red rag simply by looking at its haircut. Cows have a facial hair whorl, a twist of hair that grows either high on their foreheads or below the eye line, with around one in ten cattle forgoing this fashion statement altogether. Bulls that either carry this whorl high on their heads or do not have one at all are much more likely to give matadors and farmers a run for their money. They become agitated far more quickly and are more likely to head-butt people or ram fences or walls.[7]

Tardigrades are cuddly eight-legged animals which have rightly been given the nickname water bears. But they are also the hardiest creatures on earth. The tiny microscopic organisms, up to 1.2 millimetres long, are capable of withstanding the most extreme environments by dehydrating and going into a state of frozen animation. Some species, which live everywhere from moist mosses to the deep sea, can withstand temperatures as low as -253°C, just 20°C above absolute zero, while there is anecdotal evidence that some can survive being immersed in liquid helium, at -272°C, or just 1°C above absolute zero. They can withstand being boiled in water, being thrown into pure alcohol and a pressure of 600 megapascals, equivalent to six times the pressure of sea water at a depth of 10,000 metres. Even bacteria cannot survive in conditions anywhere approaching that extremity.[8]

There is a migrating bird that travels thousands of kilometres with little or no sleep. **White-crowned sparrows** (*Zonotrichia leucophrys gambelii*) spend 63 per cent less time sleeping during their migrating season. But despite this lack of sleep, the birds remain alert and perform as well as at other times. The birds do not even use the trick of resting one half of their brain while the other half remains awake, as some ducks, dolphins and seals do. In other animals, sleep deprivation quickly leads to illness and impaired performance.[9]

Whales get the bends. Despite regularly diving to extreme depths on the hunt for squid and other prey, **sperm whales** (*Physeter catadon*) can suffer decompression sickness, something not thought possible for over a century. The bends, caused by nitrogen gas bubbling out of the bloodstream as the whale surfaces from deep waters at great pressure, is a progressive condition that affects the whales over their whole lives, causing pits and erosions to form on the whale's nose and rib bones.

The long-extinct **plesiosaur**, a giant marine reptile, also suffered from the bends.[10]

One species of spider manages to spend its entire life living underwater. Using a dense mat of specialised hairs that cover its body and abdomen, the **water spider** (*Argyroneta aquatica*) traps a bubble of air around its body. The spider spends its whole life within this bubble and even breathes the trapped air.[11]

Some species of freshwater **turtle** spend months at a time continually submerged underwater, hibernating during winter when temperatures fall. There is no conclusive evidence, however, that sea turtles do the same.[12]

Hagfish (*Eptatretus stoutii*) produce substantial amounts of slime when harassed. The slime deters predatory fish by clogging their gills, potentially suffocating them. That may explain why hagfish are rarely eaten by other fish, and instead tend to be eaten by seabirds, seals and sea lions and toothed whales and dolphins.[13]

Good vibrations

THERE ARE LARGE numbers of anecdotal reports that animals can predict earthquakes, but ants at least cannot. Researchers had the serendipitous opportunity to answer the question once and for all when the ground began to tremble in the Mojave Desert, California, at 5.58 a.m. on 28 June 1992, right in the middle of a study on the behaviour of **the desert harvester ant** (*Messor pergandei*). Measurements of how many ants walked the trails to and fro from the colony, the distributions of worker ants, and how much carbon dioxide they produced as they worked, revealed that the ants did not change their behaviour one jot before, during or after the 7.4 magnitude earthquake, and were unfussed by the largest quake to strike the USA in four decades.[1]

Termites do a spot of headbanging to warn each other that the colony is under attack. When a nest of the **dampwood termite** (*Zooteropsis angusticollis*) becomes infiltrated by a deadly fungus, infected individuals begin waggling their heads repeatedly, which sends a seismic signal through the nest, giving uninfected termites a chance to escape.[2]

The **southern green stink bug** (*Nezara viridula*) has a special way of finding a partner. For a start, it is the girls that woo the boys, and they do it by posting the insect equivalent of a road map to love. Females vibrate their bodies at a frequency close to 100 hertz, sending the vibrations down through the leaves and stems of a plant. Alighting males detect these vibrations through their feet and antenna. Remarkably the vibrations are so precise that the males can use them to beat a direct path to their mate, some-how taking every right turn at every plant stem junction.[3]

How does a mole rat find its way underground without being able to see, hear and smell beyond the earth in front on its face? The **blind mole rat** (*Spalax ehrenbergi*) uses seismic echolocation, or in other words, is a bit of a headbanger. It bangs its head against the walls of the tunnel it has created, and then, using tiny mechanore-ceptors in its feet, listens out for any echoes as this seismic signal reverberates underground. The vibrations picked up by its paws paint a clear picture of what lies ahead, allowing the mole rat to make detours efficiently around any objects in its way, such as rocks.[4]

American lobsters (*Homarus americanus*) have what may be a unique ability among crustaceans. The lobsters emit sounds into the ocean at frequencies ranging from 87 to 261 hertz by con-tracting their internal muscles and vibrating a part of their exoskeleton called the carapace. Why the lobsters do this, or what function the sounds serve, is unknown.[5]

Questions, questions, questions

∾

Why do zebras have stripes?

A ZEBRA'S STRIPES DO not create an optical illusion that makes the animal look larger, do not dazzle or confuse predators, or make it difficult for predators to spot an individual within a herd. Nor do they help camouflage zebras within long grass, nor provide a unique pattern that animals can use to identify each other. The most likely explanation is that a zebra's stripes offer it protection against biting insects, in particular the **tsetse fly** (*Glossina* spp.).

The three species of zebra, **plains zebra** (*Equus burchelli*), **mountain zebra** (*Equus zebra*) and **Grevy's zebra** (*Equus grevyi*) each have a distinctive pattern of stripes. For instance Grevy's zebra have narrower stripes than plains zebra. The quagga, a subspecies of the plains zebra once common in South Africa but which became extinct around 100 years ago, had stripes only on its front half, its rear being brown in colour. No one knows why each sports a different pattern.[1]

Why are egg yolks yellow?

Eggs yolks are yellow because they contain yellow pigments known as carotenoids that help give the developing chick a head start in life. Carotenoids are antioxidants, and mother birds enrich their eggs with the chemicals to help protect the embryo and growing chick from free radicals produced by the high rates of metabolism needed to grow fast. Free radicals are toxic and damage DNA, lipids and proteins in the growing chick. Higher quality mothers can add more carotenoids, helping to produce higher quality offspring. Egg producers also add carotenoids to the feed of **chickens** so they produce brighter-coloured egg yolks, which consumers prefer.[2]

Why do bird chicks have brightly coloured gapes?

A chick with a bright red gape is advertising to its parent that it is in rude health, as the redder the gape, the more red carotenoid pigments it has that can be utilised by its immune system to fight off parasites. Parent **birds** respond to this signal by giving more food to chicks with brighter gapes.[3]

How do songbirds sing their songs?

All **songbirds** have an organ called the syrinx, which is located deep within the thoracic cavity at the base of the trachea. Sounds produced by the syrinx then travel up through the vocal tract before exiting through the beak. By moving the vocal tract, and the size of the beak gape, birds can produce a variety of whistles and tunes. There is also a relationship between the size and shape of a bird's beak and the type of song it sings. In Darwin's finches, for example, birds with large beaks produce songs with comparatively low trill rates and narrow frequency bandwidths, whereas birds with small beaks produce songs with fast trill rates and wide frequency bandwidths.[4]

How much effort does it take for a bird to sing?

The longer a **Carolina wren** (*Thryothorus ludovicianus*), **European starling** (*Sturnus vulgaris*) or **zebra finch** (*Taeniopygia guttata*) sings, the higher is its metabolic rate, which eats up valuable energy reserves. Nightingales also lose more of their body weight overnight the more they sing. That is why many birds sing along to a dawn chorus – to display how fit they are in evolutionary terms. They are showing potential mates that they can afford to waste energy singing at a time when they could be sleeping. At dawn, **nightingales** (*Luscinia megarhynchos*) sing 10 separate songs every minute. Crowing cockerels sing with less acoustic power in relation to their weight than either **song thrushes** (*Turdus philomelos*) or **European robins** (*Erithacus rubecula*).[5]

Why do honeybees dance?

This elaborate communication ritual, where returning **honeybees** (*Apis mellifera*) waggle their behinds to tell other nest mates about the whereabouts of food, is thought to be an adaptation to warm, tropical climates. In colder, temperate climes, bees feed on widely distributed herbs and shrubs, so it is relatively easy for individuals to find their own sources of food. In tropical forests bees feed on flowering trees, which are few and far between and only flower for a few days at a time. One bee will have little chance of locating its own food and therefore relies on others in the colony to pass on directions to the few sources of nectar nearby.[6]

How does a bumblebee first find a flower?

(Especially if a bee has never seen a flower before, or has never had a chance to learn from another bee how to identity a productive source of food.)

It turns out that **bumblebees** (*Bombus impatiens*) have an innate preference for flower-like shapes and patterns, and are also naturally attracted to the colours blue and yellow, which are commonly used by flowers from which the bees gather nectar.[7]

Why do many mammals eat the afterbirth of their offspring?

One reason is because it contains painkillers that help bring on maternal feelings and ease the distress of having just given birth. Afterbirths and amniotic fluid contain a chemical known as Placental Opioid-Enhancing Factor or POEF, which affects how the brain uses opioids to numb the sensations of pain coursing through the nervous system. It is thought that females produce this chemical when pregnant, which dulls the pain of giving birth. The mothers then eat the afterbirth, both to get another hit of the drug, but also because the chemical appears to trigger maternal behaviour towards the newborn.[8]

Why does the hammerhead shark have a head shaped like a hammer?

Having the eyes far apart does not help the **hammerhead shark** (*Sphyrna lewini*) see any better, and the hammer shape does not aid manoeuvrability. Nor, as once thought, is it used to bash the seafloor in a bid to dig up bottom-dwelling fish or shrimp upon which the shark regularly feeds. Rather the wide shape of the

hammerhead helps the shark sweep a greater area of the sea floor with electroreceptive organs that sense the electric fields of its prey.[9]

Why do kangaroos and wallabies thump their feet?

A few reasons have been proposed as to why kangaroos and wallabies thump their hind feet on the ground. One is to warn others that a predator such as a dingo might be approaching. As well as creating a loud noise, the force of the thump on the ground creates a seismic vibration that may be detectable by another kangaroo some distance away. However, solitary kangaroos thump their feet significantly more often than kangaroos living in groups of two or more. It is more likely that the thump itself acts as a signal to an encroaching predator, either to startle it or warn that it has been spotted by the kangaroo.[10]

Why do bats prefer to roost in old churches, warehouses, lofts and other buildings?

It is a pertinent question considering there are perfectly adequate natural roosts all around. The answer, it appears, is that man-made constructions provide an all-round better home. Any roost used by female bats must provide a safe and secure environment that protects them from predators and adverse weather conditions, as well as providing a microclimate conducive to the growth of their offspring while in the womb and after they are born. It must also be a space with design features that allow the bats to interact socially.

Studies of maternity colonies of **big brown bats** (*Eptesicus fuscus*) show that bats living in buildings enter a state of torpor, or deep sleep, less often than bats living in rocky outcrops or caves. Entering torpor less frequently allows the bats to be more active on more days, giving them increased time to grow and develop their foetus. Bats in building roosts give birth earlier than those in rock roosts.

Bats in buildings also save more energy than rock-roosting individuals by roosting in the warmer microenvironments of buildings. This allows females to achieve higher body temperatures during the day, and juveniles to achieve higher body temperatures during the night when females are away foraging. A warmer building roost helps juveniles grow more quickly and young building-roosting bats fledge one to two weeks before rock-roosting bats. In short, relative to natural roosts, man-made buildings help protect bats against predators, and help them save energy, allowing the bats to give birth earlier and their young to grow faster.[11]

What is the best design for a spider's web?

The simple answer is that no one yet knows for sure, and each web design works well for the species that weaved it. But it does appear that the huge, glorious symmetrical webs spun by orb weaving **spiders** are actually the most primitive type of web, and many species of spider have rejected this design in favour of more advanced types. Orb webs, which as their name suggests are often large and circular, are usually strung between leaves and branches, and are designed to catch winged prey such as moths or flies.

Many spiders moved on to evolve sheet webs, which, as their name suggests, are relatively flat, innocuous matted webs that are strung across places such as leaf litter and dark hollows where prey are likely to live. These sheet webs may be designed to catch jumping and flying insects taking off from the ground.

Sheet webs do not have the beautiful radial structure of orb webs, but they do come in a variety of designs. Some, for instance, may concave up and are known as domed sheet webs. Others concave down, and are called bowl sheet webs. Many species of spider prefer to weave domed webs rather than bowl webs, which is odd as it is much easier to produce a bowl-shaped web under tension.

Finally, sheet webs evolved into what are called gumfoot webs, where many of the sticky lines of silk are attached not to each other but to the ground. The web then almost completely fills an enclosed space.

One reason why webs evolved from large simple orbs into smaller, more complex gum foot webs may have been to give spiders more protection. Orb spiders sitting in the middle of an orb web are relatively easy to see, and can be easily picked off and eaten by flying predators. As more and more spiders fell victim to agile predatory flying insects, the arachnids gradually fought back by evolving smaller webs that offer more protection.

Of the more than 10,000 species of web-weaving spiders known as *Orbiculariae*, 60 per cent have forgone the orb web design in favour of spinning webs of a different shape.[12]

Many species of orb web spider also adorn their webs with intricate patterns and stylistic flourishes known as silk decorations. These can take many forms; for example the **spider** (*Argiope versicolor*) will weave Z-like patterns of silk radiating out from the centre, whereas another species, *Araneus eburnus*, weaves similar patterns in a straight line, say from top to bottom of the web. *Octonoba sybiotides* creates a beautiful spiral within the centre of its web, while *Gasteracantha minax* adds little tufts of silk all over the web structure. But the exact reason why the spiders do it remains controversial.

The rationale behind silk decorations is further complicated by the fact that some species add different decorations depending on whether they are adults or juveniles, whereas individuals within a single species might decorate their webs in different ways. A single individual spider will even decorate its web differently from day to day.

One idea is that these silk decorations act as anti-predator devices. **Decorating** spiders (*Argiope trifasciata*) are more likely to be taken by predatory wasps than their non-decorating counterparts, while there is evidence that juvenile spiders of the species A. *versicolor* run to the other side of the web when attacked, using the decorations as a physical barrier. But other evidence suggests that decorations can actually attract some predators. The predatory **praying mantid** (*Archimantis latystylus*) is attracted to decorated webs made by the **spider** (*Argiope keyserlingi*), regardless of whether a spider is actually at home or not.

Another theory is that silk decorations help attract prey for the spiders to feast on. It does appear that many species of fly are attracted to the decorations weaved onto some spiders' webs, and it could be that the decorations help reflect ultraviolet light, either appearing to the flies to be gaps between leaves, or mimicking the ultraviolet light signal produced by plants to encourage insect pollinators. However, spiders that have a full stomach tend to produce more decorations than spiders that are hungry.

The final idea is that spiders decorate their webs to help protect them. The added silk may offer some mechanical support to the web, or it might serve to warn passing animals such as birds of the web's presence, ensuring they do not walk or fly straight into it, destroying the spider's hard work.[13]

Enduring enigmas

◆

W̲ᴇ ꜱᴛɪʟʟ ᴅᴏ ɴᴏᴛ know why male lions have large dark manes – the only cats to do so. A number of ideas have been postulated. One long-standing notion is that the mane helps protect males when they are fighting one another, and evolved to shield the neck and shoulders from the lethal claws and teeth of other **lions** (*Panthera leo*). However, studies have shown that when male lions fight, they do not specifically attack the area of the body covered by the mane. Injuries sustained in the mane area are no more severe, or more likely to kill a lion, than injuries suffered elsewhere on the body. Regions of the mane that are injured do not correlate with those parts of the body where a mane develops first in young adults, or are covered by longer, darker hair in older more dominant males, which would be expected if the mane acted as a protective shield against teeth or claws.

A more likely idea is that a mane acts as a sexual advertisement, both to attract females and scare off other less dominant males. Males with longer and darker manes tend to win more fights with rivals. Males with dark manes, which indicate higher testosterone and nutrition levels, enjoy longer reproductive life-spans while their offspring have a higher rate of survival. Castrated males and those that have been badly injured also lose their manes, suggesting hormones play a significant role in their growth.

It is also likely that climate plays a role in the development of the lion's mane, and that growing a mane comes at some cost to the male. Maned males tend to have higher body temperatures than females, and males tend to grow lighter and shorter manes in hotter seasons, years, and habitats. A study of lions living in zoos in the USA also shows that lions that endure colder temperatures during the year have thicker, more extensive manes, and also grow hair on their ribs and bellies. Overall, climate variation may account for up to half of the differences seen in mane growth.[1]

No one knows exactly how **fireflies** generate their stunning flashes of bright light which they use to attract mates.[2]

Exactly how **bats** grew their wings remains a mystery, as no fossil exists which shows an animal somewhere between a modern bat and the wingless mammalian ancestor it came from. Winged bats just suddenly appeared about 50 million years ago.

There is evidence that the evolution of bat wings was triggered by a mutation in a single gene called BMP2, one of a family of genes that play an important role in the development of mammal limbs. When a protein produced by this gene is applied to the digits of an embryonic mouse growing in the laboratory, the mouse will develop hugely elongated fingers, just like the long digits that hold the membrane that forms a bat's wing.[3]

No one knows why some animals, such as many species of monkey and ape, have evolved fingernails. One idea is that nails evolved from claws to help climbing mammals, such as the earliest primates, to grasp branches and tree trunks better. However, there is no conclusive evidence that this is true. For example, one species of primate, the claw-bearing **midas tamarin** (*Saguinus midas*) has just as good grasping ability as the nail-bearing **South American squirrel monkey** (*Saimiri sciureus*). There is only one mammal, apart from the primates, that has fingernails which it uses for climbing. The tiny 7–16 gram **honey possum** (*Tarsipes rostratus*) has fully reduced its claws into nails, just as many species of monkey and ape have. One group of New World monkeys, the

Callitrichinae, has re-evolved functional claws from fingernails, structures which themselves evolved from ancestral claws.[4]

The **Chinese water deer** (*Hydropotes inermis*) is the only species of deer, or cervid, that does not grow antlers.[5]

Wandering albatrosses (*Diomedea exulans*) are able to pinpoint the specific remote island where their nests are located after making foraging flights of several thousands of kilometres over featureless ocean. They do not rely on the earth's magnetic field, and no one knows how the birds acquire such precise and impressive navigation ability.[6]

Mysterious giants

❧

In the depths of the African jungle roams a mysterious giant ape. For the past few years, no one knew if this ape was an out-sized chimp, far larger than any previously recorded, a hitherto unknown hybrid between a chimp and gorilla, or an entirely new species.

The legend of these apes has abounded for over a century, with local people living in the northern dense forests of the Belgian Congo, now the Democratic Republic of Congo (DRC), describing them as giant ferocious lion killers. The rumours were given fresh impetus in 1898 when a Belgian army officer returned from the region with three large ape skulls, which have since been identified as gorilla skulls. Similar skulls have also turned up more recently in the region, but the nearest gorillas live over 500 kilometres away.

The mystery deepened when it was realised that although each skull had a prominent sagittal crest, a bony ridge running along the length of a typical male gorilla's skull, the animals that the locals described appear to behave more like chimps. Whereas gorillas would charge approaching hunters, these beasts quietly disappeared back into the trees as chimps do.

In 2001, scientists undertook an expedition to the DRC to try to solve the enigma. Instead, they became even more confused. They found evidence of nests built on the ground by huge ape-like creatures, reminiscent of gorilla nests rather than chimp nests, which are usually built in tree branches. Snatched photographs showed apes that had gorilla-like faces, yet were grey all over rather than sporting grey silverbacks like gorillas.

However, DNA evidence of hair and faeces, and a closer

examination of skulls recently found in the region, suggest they come from chimps, although the skulls are far larger than any chimp skulls recorded so far, as were footprints found on the forest floor. Experts also estimate from photographs that these large apes grow up to 2 metres tall and weigh half as much again as the largest recorded chimps.

In 2006, researchers also announced they had tracked these huge beasts for over 20 hours through the jungle north-west of Bili in the DRC. DNA and behavioural evidence further suggest the beasts are chimps of the species *Pan troglodytes schweinfurthii*, and, as suspected, there is evidence that the apes have a unique culture, nesting on the ground, using exceptionally long sticks as tools to pick up specific species of ant and using the shells of turtles to smash open termite mounds. What is more, the lore that the beasts are lion killers has a ring of truth, as there is evidence that the outsized chimps regularly feast on a diet of mammal flesh.[1]

For years, strange, eerie, deep sounds have been recorded booming underwater across the world's oceans. And some biologists suspect they may be produced by, as yet, unknown sea creatures. One such sound, nicknamed Bloop by researchers in 1997, was a rapidly changing, low-frequency sound similar to that produced by large whales. But it was recorded by underwater microphones over 4,800 kilometres apart on opposite sides of the ocean, much further apart than any animal sound so far known could travel, meaning it should also have been louder than any known animal noise. Some biologists suspect the sound was produced by whales, others by squid, while others suggest it was emitted by natural inanimate objects such as ice sheets breaking up. Others still suspect is could be made by an extremely large creature yet to be discovered.[2]

A similarly mysterious ocean sound has been heard in the oceans surrounding Hawaii, and has been recorded by US naval submarines. Dubbed Boing, the noise sounds like a sort of fluttering echo, and perplexed all who heard it. Some thought it could be made by a large fish, others by enemy submarines. In 2005, however, the mystery was solved when biologists discovered it was actually a call made by breeding **minke whales** (*Balaenoptera acutorostrata*).[3]

Huge, strange gelatinous blobs regularly wash up on beaches around the world, sparking speculation that they are the carcasses and remains of mysterious, giant sea creatures, such as giant octopi, squid and other undiscovered species.

Famous examples include the 'giant octopus of St Augustine' from Florida in 1896, the 1960 Tasmanian west-coast monster, two Bermuda blobs from the 1990s, the 1996 Nantucket blob, and an unidentified creature that washed ashore in Newfoundland in 2001. All, it turns out, are just the decomposed blubber of the **sperm whale** (*Physeter catadon*).

The most recent incident occurred off the coast of Los Muermos, Chile, in July 2003, when a 13-tonne mass of amorphous tissue rolled ashore. Local marine researchers could find no bones within the blob, leading to speculation it was the leftovers of a hitherto unknown giant octopus that patrolled the ocean deep. But it was only with the discovery within the gelatinous mass of collagen fibres similar to those found in whale blubber, sperm whale skin ducts and whale DNA, that it was confirmed the Chilean blob and others were nothing more that the rotting leftovers of a long-dead leviathan, and the mystery finally solved.[4]

In 2004, the first ever giant squid was caught breaching the ocean surface and it was the largest squid ever recorded. The **colossal** squid (*Mesonychoteuthis hamiltoni*), caught in the Ross Sea near Antarctica, had a mantle length of 2.5 metres, yet it was still a juvenile. From this specimen, experts infer that adult colossal squid may grow a mantle as large as 4 metres long. Add to that its giant long arms and it is likely that 15-metre-long squid are swimming the ocean depths.

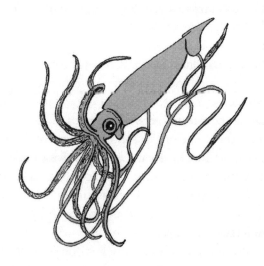

Giant squid are thought to be true monsters of the deep, reaching their extraordinary size in just a few years. That makes them one of the fastest-growing animals on the planet. The colossal squid has a powerful muscular fin and grows rotating hooks along its arms, making it a formidable adversary for any predators. The discovery of M. *hamiltoni* has also confirmed that we have for centuries suspected the wrong creature of being the Kraken. Contrary to popular belief, this beast of lore is not the **giant squid** known as *Architeuthis dux*, a massive animal that can grow to a mantle length of 2.25 metres and a total length of 13 metres. A. *dux*, while big, is actually quite a fragile animal with long, thin arms that are built more for catching smaller, quicker prey than wrestling anything as large as a whale or boat.

It is also a myth that sperm whales are the only animals capable of taking on giant squid. Sleeper sharks, one of the largest species of carnivorous shark that can grow up to 7 metres in length, have been found with remains of giant squid in their stomachs. Both A. *dux* and its even larger cousin, the colossal squid, appear to be on the sleeper shark menu, with the colossal squid the main dish of choice. The lengths of squid beaks recovered suggest that these sharks eat squid up to at least 12 metres long.[5]

A mysterious whale that no one has yet been able to identify has been swimming the Pacific Ocean since 1992. The whale sings a tune like no other, emitting a song at around 52 hertz every autumn and winter. Its calls are unlike those of any known baleen whale, such as humpback, blue or fin whales, and the path of the animal does not match the known migrating routes of any other species.[6]

The two largest whale species, the **blue whale** (*Balaenoptera musculus*) and the **fin whale** (*B. physalus*) often hybridise, producing anomalous whales that are hard to identify. Male blue and fin whale hybrids are often sterile.[7]

Man and beast don't get along

∾

As the presence of humans can have profound effects on the animals they are visiting, ecotourism is not all it is cracked up to be for some of them. For example, when people visit **yellow-eyed penguins** (*Megadyptes antipodes*), the birds delay returning to their nests after a foraging trip, which in turns affects how much food they bring back for their chicks. Fledglings living in areas where high numbers of tourists visit are lower in weight than those living in unvisited places and it is likely this has a major impact on their survival.[1]

The heart rate of a wild **emperor penguin** (*Eudyptes schlegeli*) increases 1.7-fold whenever a human approaches to within 5 metres. It also becomes six times more vigilant. A vigilant penguin displays nervous behaviours including wing shivering, swallowing and beak dipping.[2]

And the problem extends from the ice wastelands to the depths of the jungle. Chicks of the bizarre **hoatzin bird**, which sports claws on its wings and lives in the tropical rainforest, are similarly affected, with less surviving in areas where tourists visit.[3] Polar bears, dolphins and dingoes all become disturbed by ecotourists, displaying odd behaviours, such as becoming overly vigilant, that affect their reproduction.

On occasions, however, the opposite is true. When people go to the coast of British Columbia and Alaska to watch **brown bears** (*Ursus arctos*) feeding on salmon, their presence scares away the larger, dominant male grizzlies. And in their absence, female bears that are raising cubs feel more comfortable feeding and end up catching more fish both for themselves and their family.[4]

Elephants are aware of when they are living in protected conservation areas, and will go out of their way to avoid poachers and landowners with whom they might come into conflict. Herds of **African elephants** (*Loxodonta africana*) living in Kenya tend to meander in home zones, spending over half their time in specific parts of unfenced protected areas where they do most of their feeding. The elephants only cross unprotected areas via specific well-travelled corridors that lead to other protected areas. What is more, elephants travel along these corridors at high speed seemingly aware that they are at a greater risk of poaching or being hounded by people, and will often undertake these hazardous journeys at night to avoid being spotted.[5]

The fate of different animals that leave wildlife hospitals, which seek to rehabilitate displaced, sick, injured or orphaned wild animals and then return them to the wild, can vary widely.

In the UK, for example, 30,000 to 40,000 casualties are taken into wildlife hospitals each year, yet some species have a better chance of rehabilitation than others. Rehabilitated **barn owls** (*Tyto alba*) have a shorter life expectancy, surviving for just an average of 77.9 days, than birds that spend their whole life in the wild, which live on average for 365.7 days. The average survival time of orphaned **red fox** (*Vulpes vulpes*) cubs released at 4 to 7 months of age is just 94 days, while 70 per cent of **guillemots** (*Uria aalge*) that had been caught in oil spills die within the first two weeks after release.

One species that does do well from such hospital care is the European hedgehog. Not only is it the most common mammal admitted to British wildlife hospitals but **hedgehogs** (*Erinaceus* *europaeus*) that are treated and then kept in hospital for longer than one month have a better chance of survival than those treated and released immediately. It may be that this time spent being cared for on an animal ward helps acclimatise the hedgehog to the stress of being handled and looked after by people, or it might give the animal time to put on sufficient fat reserves and weight to help it survive the first few weeks of being released back into the wild.[6]

Pity the **bats** and **birds** that become the objects of study for zoologists. There is increasing evidence that many of the tags and radio antennae fitted to these animals to allow researchers to track their movements may be life-threatening.

Researchers once commonly used two rings to study bats. The first was a metal ring, fitted to each bat's forearm, which carries a serial number used to distinguish individual animals. Researchers also liked to fit a second, brightly coloured plastic ring so they could identify different bats from far away. But these rings clink together, acting like a bell on a cat's collar. The sound is almost inaudible to most people. But unluckily for the bats, moths, one of the main items on their dinner menu, can easily hear it. Tests have shown that almost half of all moths perform an elaborate escape manoeuvre when they hear the sound, drastically limiting a bat's chance of catching any food.[7]

Penguins fare little better. The metal bands scientists fit to a penguin's flipper adversely affect the bird's ability to swim and catch fish. Banded penguins generally arrive at their breeding colonies later than unbanded penguins and go on to produce fewer chicks.[8]

Replacing the metal bands with radio transmitters may not help the penguins either. **Magellanic penguins** (*Spheniscus magellani-*

cus), living in Argentina, which were fitted with rigid antennae were five times less efficient foragers than penguins not fitted with the equipment.[9]

Darting animals with tranquillisers, in order to capture, study or allow vets to treat them, can have unforeseen and not particularly favourable consequences. Male **bighorn sheep** (*Ovis canadensis*) butt and kick each other in intense fights over females. But despite appearing to recover completely from the ordeal of being tranquillised, those sheep darted with either ketamine or xylazine soon lose their fighting skills. Sheep that have been previously darted lose more fights than before and slip down the dominance hierarchy, being overtaken by previously subordinate rivals.[10]

Animals often exhibit strange behaviours due to man-made pollution in the environment. **Herring gulls** (*Larus argentatus*) that ingest lead, for instance, find it difficult to keep their balance and often fall over. Male **ringed turtle doves** (*Streptopelia risoria*) that eat the chemical DDE do not court their mates and significantly reduce their nesting behaviour, whereas female **Western gulls** (*Larus occidentalis*) exposed while still in the egg to the related chemical DDT go on to form homosexual pairs as adults.

DDT exposure makes **bald eagles** (*Haliaeetus leucephalus*) build funny nests, while atrazine makes **goldfish** (*Carassius auratus*) hyperactive and **deermice** (*Peromyscus maniculatus*) aggressive, while the chemical TCDD makes **rhesus macaques** (*Macaca mulatta*) play more roughly. The chemical additives PCBs hasten the onset of puberty in **Norwegian rats** (*Rattus norvegicus*) while bisphenol A does the same to **house mice** (*Mus musculus*), whereas PCBs will retard the learning abilities of **long-tailed monkeys** (*Macaca fascicularis*).[11]

The most common large animals involved in motor vehicle collisions are **kangaroos** in Australia; camels in Saudi Arabia; and deer, moose, and bear in parts of Europe, Japan, Sweden, the USA and Canada.[12]

Each year in Europe there are 500,000 collisions between vehicles and hoofed animals such as **deer** or **moose**, injuring 30,000 people and killing 300.[13]

Across Alaska, on average one in five collisions with a **moose** (*Alces alces*) results in an injury to an occupant of the vehicle involved. One in 200 collisions results in a human fatality.

In Anchorage, Alaska, a car is 2.6 times more likely to collide with a moose when it is dark compared to during the day. During the five-year period from 1991 to 1995, vehicle collisions with moose in Anchorage cost an estimated US$ 10 million, accounting for 5 per cent of all costs associated with motor vehicle crashes. Overall, 2 per cent of all vehicle crashes in the area involved striking a moose.[14]

Turtles often fail to nest on beaches when people are nearby. The reptiles will refuse to come out of the water, will abandon their nesting attempts, or will dive back into the sea when boats approach.[15]

Larger **bird** species tend to fly away when approached by people more quickly than smaller species. Carnivorous and omnivorous species tend to be more 'flighty' than herbivorous ones.[16]

Sturgeon fish are so highly prized for their caviar that sovereigns of many countries have claimed the sole rights to the fish. Henry II of England (AD 1133–89) placed sturgeon under royal protection, and the British monarch still retains the first right to any sturgeon caught along the British coast.[17]

The sounds of the city can have a disturbing effect on animal behaviour. In central Thailand, three species of **frog** (*Microhyla butleri*, *Rana nigrovittata* and *Kaloula pulchra*) that live at the edge of ponds decrease the rate at which they call when they are drowned out by the sound of planes flying overhead or motorcycles driving by. Strangely, another species of **frog** (*Rana taipehensis*) reacts to this lull in the amphibian chorus by calling more often, probably to increase its chances of being heard among all the frogs present.[18]

Birds, particularly **blue tits** (*Parus caeruleus*), **great tits** (*P. major*) and **chaffinches** (*Fringilla coelebs*), will begin singing their dawn chorus earlier in the day when in cities than when in the country, possibly to avoid competing with the din made by people and traffic.[19]

Florida scrub jays (*Aphelocoma coerulescens*) living in the towns and cities of Florida in the USA appear to have a dream life, feeding often and well on nuts and treats put out by well-meaning local people. But this easy living comes at an unforeseen cost to the birds. The huge and artificial abundance of food fools the jays into thinking that spring has come a few weeks early, and the city birds breed earlier and lay larger clutches than their country cousins. On the surface this appears to be no bad thing, but while the nutritional content of the nuts and other plant-based foods fattens up the adults, it is not suitable for the young nestlings the jays are feeding. Chicks require protein from grubs and other insect larvae that usually appear later in the year. By eating the human-provided food, the nestlings become malnourished and either suffer from stunted growth or starve to death.[20]

One bird of prey species also fares badly in towns and cities. Large numbers of **Cooper's hawks** (*Accipiter cooperii*) are drawn into the city of Tucson, Arizona, lured by the large bounties of urban-dwelling pigeons. Although higher densities of the hawks are found in the city than in the surrounding countryside, the city lifestyle comes at a high price. Hawks nest earlier and lay more eggs in the city, but more than half their chicks die after hatching. The cause of the problem is a disease called trichomoniasis, which is carried by town-dwelling pigeons and doves, which are a mainstay of the urban hawks' diet.[21]

Sperm whales (*Physeter catodon*) fell silent during the US invasion of Grenada in 1983. The whales stopped making their usual series of clicks and presumably stopped foraging in response to the navy's use of sonar in the surrounding waters.[22]

Differences

The following differences can be seen seen in captive animals compared to wild animals:

Lions	Shorter skulls and smaller brains
Horses	Smaller brains, separated incisors
Leopards	Broader muzzles
Alligators	Smaller and flatter skulls, larger bodies
Chimpanzees	Longer limb bones
Red grouse, geese, bustards	Shorter intestines, lighter hearts, livers and gizzards
Fruit bats	Fatter
Gophers, anteaters, bears and camels	Osteoporosis
House mice	Greater numbers of neurons in the brain, greater ability to learn
Salmon	Smaller heads, larger bodies
Silver foxes, mink and salmon	Less sexual dimorphism
Marsupial feathertail gliders	Active for longer, suffer hypothermia
Bears	Greater tooth decay

table source [1]

When animals attack

Iɴ ᴏʀᴅᴇʀ ᴏf ᴛʜᴇ ᴍᴏsᴛ to the least deadly, shark species responsible for the majority of human deaths (not including shark attacks following air–sea disasters) are:

Great white (*Carcharodon carcharias*)
Tiger (*Galeocerdo cuvier*)
Bull (*Carcharhinus leucas*)
Requiem (*Carcharhinus* spp.)
Sand tiger (*Carcharias Taurus*)
Blacktip (*Carcharhinus limbatus*)
Hammerhead (*Sphyrna* spp.)
Spinner (*Carcharhinus brevipinna*)
Blue (*Prionace glauca*)
Blacktip reef (*Carcharhinus melanopterus*)
Lemon (*Negaprion brevirostris*)
Bronze whaler (*Carcharhinus brachyurus*)
Nurse (*Ginglymostoma cirratum*)
Shortfin mako (*Isurus oxyrinchus*) [1]

One species of shark has been known to attack nuclear submarines. The **cookie-cutter shark** (*Isistius brasiliensis*) measures just 50 centimetres in length, yet has been recorded taking chunks out of the rubber sonar domes of submarines with its razor-sharp teeth.[2]

In the USA, alligators are often far more dangerous than sharks. In the state of Florida between 1948 and 2005, **American alligators** (*Alligator mississippiensis*) attacked 391 people compared to

592 people attacked by sharks. But alligators killed 17 people during this time, compared to just nine killed by sharks. That means that alligators in Florida kill 4.3 per cent of their victims, whereas sharks kill just 1.5 per cent.[3]

The huge emu-like birds, **cassowaries** (*Casuarius casuarius johnsonii*) are one of Australia's more dangerous inhabitants. The birds have lost enough fear of people that they regularly attack unsuspecting passers-by, which is unnerving considering that a mature cassowary stands up to 2 metres tall, can weigh 85 kilograms, and sports an impressive 12-centimetre spike on each foot. Cassowaries seriously injured at least six people between 1990 and 1996, with over 150 people being attacked, as well as 35 dogs, three horses and a cow, in just three areas where records have been collected in northern Queensland. Three-quarters of the human victims were attacked for the food they were carrying.[4]

Polar, brown and black bears can be dangerous, but the **sloth bear** (*Ursus ursinus*) of Asia belies its name. It is much more likely to attack than flee, and, in one region of India, over a recent six-year period, was recorded as having attacked 745 people, killing 54. Many people in the region consider sloth bears as more dangerous than tigers.

In contrast, there has been only a single recorded fatality caused by **European brown bears** (*U. arctos*) in Scandinavia in the past century, although attacks are more common in eastern European countries such as Romania. Brown bears in Siberia and North America generally go out of their way to avoid people, although attacks do happen.

Black bears (*U. americanus*) killed just 40 people in North America in the 20th century, whereas **polar bears** (*U. maritimus*) encounter so few people that fatalities are rare.[5]

In India, a **tiger** has to kill at least three people before it can be considered a man-eater. The Sundarban Tiger Reserve in the Ganges delta is the only place on earth where man-eating tigers regularly abound. It is thought this may be because the reserve is the only place on the Indian subcontinent where tigers have never been hunted for sport, and so have never developed a healthy fear of humans.[6]

Workers milking venom from **lance-headed vipers** (*Bothrops moojeni*) on a snake farm in Brazil will expect to get bitten 2.73 times every 10,000 days of work, and 3.51 times for every 100,000 times they attempt to extract venom.[7]

Only four groups of spider are capable of killing people with a venomous bite. They are the **widow** and **redback spiders** (*Latrodectus* spp.), **recluse spiders** (*Loxosceles* spp.), **Australian funnel-web** spiders (*Atrax and Hadronyche* spp.), and the **Brazilian armed spider** (*Phoneutria nigriventer*).[8]

In Australia:

Twenty-eight per cent of people attacked by **redback spiders** (*Latrodectus* spp.) are bitten while putting on their shoes.

Forty-eight per cent of people attacked by **cupboard spiders** (*Steatoda*) are bitten while getting dressed.

Seventy-six per cent of people attacked by **huntsman spiders** (*Sparassidae*) are bitten while interfering with the spider.

Sixty-three per cent of people attacked by **white-tail spiders** (*Lamponidae*) are bitten while the spider is trapped between their clothes and skin.

Forty-two per cent of people attacked by Australian funnelweb, mouse, and **trapdoor** spiders (*Mygalomorphae*) are bitten while in the garden.[9]

In Brazil:

In a survey conducted in Pernambuco, Brazil, over 48 per cent of the people attacked by a poisonous animal were bitten by an **arachnid**. Just over 40 per cent were bitten by snakes, with just 4 per cent by insects. Over 2 per cent were assaulted by venomous miriapodes, a class of animal that includes centipedes and millipedes, and just over 1 per cent were attacked by aquatic animals. The attacks took place at a similar rate whatever the time of year, although snakes were twice as likely to bite during the day as at night, and bites were twice as frequent in rural areas than urban ones. Almost all arachnid incidents involved scorpions (over 98 per cent) taking place predominantly in urban areas (90 per cent). Over 36 per cent occurred during the night.[10]

Vampire bats (*Desmodus rotundus* and *Diaemus young*) have been known to attack people, and kill them by imparting the deadly disease rabies. In just a few months in 2005 they were responsible for biting 1,300 people in Brazil, killing 23. The attacks all took place at night-time in the northern Brazilian state of Maranhão, and of the 23 people that died 18 were children, most likely because they sleep more soundly at night and are less likely to feel the bat biting into their skin to feed on their blood.

Vampire bats usually prey on cattle and horses, but are thought to have entered the homes of people in Brazil through open windows and between gaps in the floor and ceiling. The problem is likely to worsen as deforestation means more people are living closer to bat colonies, and increasing numbers of livestock are encouraging the bats to develop larger colonies.[11]

We used to think that, with the odd exception, **snakes** were the only venomous reptiles. That may not be so. It now appears that two major groups of lizards, the monitor lizards and iguanians, which include iguanas and chameleons, are also venomous.

Previously, only two venomous lizards were known, the **Gila monster** (*Heloderma suspectum*) and the **Mexican beaded lizard** (*H. horridum*). Both were thought to have evolved their danger-ous toxins independently from snakes. In 2005, however, it was discovered that monitor lizards and iguanas also produce nine of the same toxins made by advanced venomous snakes, and other toxins that have not been identified before. And this means that pet owners need to watch out. Many people keep pet lizards, and when they are bitten, people often assume that a wound is fester-ing due to bacteria within the lizard's saliva. It might be that venom is to blame, as one hugely popular pet, the **bearded dragon**

(*Pogona* spp.), has been discovered to make the same classic toxins as rattlesnakes.[12]

It might also be that we can no longer divide up **snakes** into dangerous venomous species, and non-venomous species that pose little risk to people. It is being increasingly noticed that snakes do not necessarily need to possess fangs to be venomous. Experts had assumed that snakes that lack fangs also lack venom, but close examination of many supposedly harmless species of snake is revealing that more and more are actually venomous to a degree.[13]

Nature red in tooth and claw

❧

ALTHOUGH PENGUINS are often portrayed to be paragons of virtue, adult **emperor penguins** (*Aptenodytes forsteri*) will attempt to kidnap the chicks of another breeding pair. They will forcibly wrestle the juvenile away from its parents, while the biological parents try to protect the chick by fighting back, by pecking or using their flippers to force the intruder away.

Kidnapping behaviour often occurs when a penguin that has failed to breed sees a chick begging its parents to be fed, and interprets the juvenile's behaviour as a cue to parent it. But no one really knows why this happens, as the abducting penguin appears to benefit little from its actions – in most cases kidnapping and adoption only last for a few hours. Yet the kidnapped juveniles do suffer, as they are seldom fed and sometimes die during the struggle. Parents rarely readopt their chicks, and will only do so when their offspring have been rapidly abandoned by the kidnapper, when the parent is still close to the chick and when the abandoned chick calls its parent.

One idea is that penguins try to kidnap chicks because they are feeling a little hormonal. Female emperor penguins lay their eggs then leave to spend two months at sea searching for food before returning to feed their chicks. Those penguins that have successfully bred have lower levels of a hormone called prolactin, which is thought to stimulate the initial parenting behaviour. Females that have failed to breed are thought to maintain high levels of this hormone even when they leave the colony. When they return this high level of hormone creates a strong maternal drive that causes them to seek out and kidnap a chick of their own.[1]

A female **house sparrow** (*Passer domesticus*) will often seek out the nest of another female that her partner has also mated with. She will then kill the first female's young, to remove the competition and ensure that the male spends as much time as possible helping to raise her chicks.[2]

One species of fish hunts and eats its prey on land. When the **eel catfish** (*Channallabes apus*), which usually lives in the muddy swamps of Africa, spots a tasty insect meal, it will propel itself out of the water. It lifts the front part of its body, bending its head and mouth downwards until it makes contact with its prey, then opens and closes its mouth repeatedly until it can swallow its meal. The catfish has a specially adapted spine, which ensures the fish does not continue to nudge the prey away from its mouth, and allows it to perform this trick. It also allows the fish to hunt on land without out the need for strong weight-bearing pectoral fins.[3]

Snakes usually eat frogs and toads. But one species of toad appears to be exacting a revenge on its natural enemy. The **cane toad** (*Bufo marinus*) was first introduced into Australia in 1937, and has been going on the rampage ever since, spreading north, west and south from Queensland. This toad is also toxic, secreting a poison through its skin that can be deadly to snakes. Researchers have calculated that at least 7 species of snake in Australia would die if they ate just a single cane toad and that 30 per cent of Australian snake species will be at risk from cane toad poisoning by 2030, when the invasive anuran is expected to have conquered much of the continent.[4]

The tadpoles of the **frog** species *Rana pirica* have a unique way of avoiding being eaten. When they realise there are predatory larvae of the **salamander** (*Hynobius retardatus*) nearby, the tadpoles grow a different body shape. Within days they take on a bulging shape that is too wide to fit into the salamander larvae's mouth. As soon as the predators disappear the tadpoles grow back to their normal size.[5]

Unfortunately for the tadpoles, the **salamander** larvae also have a way of fighting back. In an incredible arms race, the salamander larvae also develop into a different shape when tadpoles are near. The beat of the tadpoles' tails is enough to cause the *Hynobius retardatus* larvae to change into a broad-headed predator morph, which is more aggressive and has a wide enough gape to gulp down their prey.[6]

Red-eyed tree frog tadpoles (*Agalychnis callidryas*) will hatch early from spawn in order to avoid being eaten by a hungry predator. When a predatory snake attacks a bundle of spawn, the tadpoles sense the vibrations, and quickly hatch in order to escape, even if they have to emerge up to three days earlier than they would do normally.[7]

Some insects such as the **praying mantid** (*Parasphendale agrionina*) can literally feel a predatory bat approaching. The insects possess hairy organs called the cercal sensory system that allow them to detect the wind generated by a flying bat's wings. A praying mantid, for instance, can detect an approaching bat around 75 milliseconds before it strikes, which usually means the bat is some 14 to 40 centimetres away. Because of the time it takes the mantid to process this information, it has just 36 milliseconds to take evasive action and avoid being eaten.[8]

Ants are such formidable predators that around 100,000 other species of insect have evolved mechanisms to coexist with them, rather than be hunted and eaten by them. Adaptations include evolving armour designed to resist attack from ants, mimicking other animals or plants to avoid being detected and secreting sugary delights such as honeydew to feed or appease the voracious ants.

About 10,000 species, including butterflies, crickets, beetles and flies go a step further and actually live within ant colonies where they find safety and rich sources of food. One butterfly species, **rebel's large blue** (*Maculinea rebeli*), has come up with a neat way to ingratiate itself into an ants' nest. The caterpillar of the large blue settles beneath its food plant to await discovery by **red ants** (*Myrmica* spp.). By secreting hydrocarbons that mimic those made by *Myrmica*, the caterpillar tricks a foraging worker into taking it into the nest, where it is placed among the ant grubs. The caterpillar then moves to safer chambers, returning periodically to binge-feed on ant grubs.[9]

Some ant species make slaves of others. **Ants** within the subfamily *Formicinae* will go out and raid the nests of other species nearby, and steal their eggs and pupae. These are taken home, where the resulting young are raised as slaves, having to do all the foraging, cleaning and babysitting for their masters.[10]

Certain species of ant jump out of trees to escape being eaten. When the **canopy ant** (*Cephalotes atratus*) is approached by a potential predator it throws itself into the air, plummeting to the ground below. But they do not free fall. By orientating their bodies, the ants steer themselves into a steep glide and head straight for the lower reaches of the tree trunk. Smaller ants tend to be more successful at hitting the trunk than larger ants, but on average 85 per cent of all ants that take a leap successfully land back on the tree. Most then walk back up to the exact branch from which they jumped.[11]

Cuckoos seem to have met their match in a bird called the **black-cap** (*Sylvia atricapilla*). Blackcaps can distinguish the eggs of the **common cuckoo** (*Cuculus canorus*) from its own, and consistently reject those laid by a cuckoo in its nest. Blackcaps lay eggs of a very similar colour and size, a uniformity that may enable them to spot the intruding egg.[12]

Wild **bonnet macaques** (*Macaca radiata*) run away faster if they see the front legs or face of a hiding leopard than if they see its back legs or hindquarters. This suggests that the monkeys have learnt to spot the rosettes, spots, and flecks of the big cat and simply react to anything that looks like a stalking predator. Conversely, they also react less quickly to stationary leopards with pure black coats, the rare melanin form, than those with conventional spotted coats, suggesting leopards benefit less from a camouflaged coat nowadays than they used to generations ago.[13]

Deep in the Kalinzu Forest in Uganda, **red-tailed monkeys** (*Cercopithecus ascanius*) hunt **green pigeons** (*Treron calva*) and then fight over the spoils with **blue monkeys** (*Cercopithecus mitis*).

It is the first example of a cercopithecoid primate, or Old World monkey, hunting birds that are outside of their nests and moving freely around the forest canopy. It is also the only place where red-tailed monkeys, which are usually thought to eat insects and fruit, have been seen hunting other vertebrate animals. Previously, only **chimpanzees** (*Pan troglodytes schweinfurthii*) were known to hunt agile vertebrates that are quicker than themselves frequently, and when chimps go hunting in this way, they usually do so by co-operating with one another and coordinating their attack. Red-tailed monkeys on the other hand have been seen stalking green pigeons until they are within just two or three metres, then pouncing on and catching the unsuspecting bird. They then pull off the feathers around the pigeon's head and begin to devour it head first. The hunt does not end there, however, as larger blue monkeys also seem to find green pigeons quite tasty. Although blue monkeys do not try to catch green pigeons themselves, possibly because they are too big and slow, they will attempt to chase red-tailed monkeys and try to poach the bird from them.[14]

To make sure they do not get eaten, **minnows** have an unusual way of protecting themselves. If one unlucky fish is attacked by a predator, specialised epidermal cells in their skin release a specific warning chemical which alerts the rest of the shoal. When minnows smell this chemical they go into hiding. Their behaviour also changes for a long time afterwards, as the minnows flee more quickly next time they see a marauding predator.[15]

The innocuous mussel can be a voracious cannibal. At certain times of the year, up to 70 per cent of all food eaten by the **green-lipped mussel** (*Perna canaliculus*) is the larvae of its own species.[16]

Poisonous spiders inject more venom into difficult or dangerous prey such as blowflies and beetles than they do innocuous victims such as crickets and stick insects. It takes each **spider** (*Cupiennius salei*) up to two weeks to replace its venom after biting, suggesting that the arachnids ration out their venom according to how difficult each prey is to overpower.[17]

Uloboridae spiders rely on an exceptional method of killing their prey. They literally wrap their victims to death. An individual small **uloborid spider** (*Philoponella vicinawill*) will weave over 140 metres of silk to wrap a single large prey item such as a moth or beetle, and will perform over 28,000 individual wrapping movements using its hind legs. It binds the silk shroud so tight that it compresses the prey's body, breaking the insect's legs, buckling its compound eyes inwards, and often killing it outright. They are the only spiders known to use silk wraps to compact and damage their prey physically instead of using silk just to restrain their victims. Most spiders also have venous glands for paralysing their prey, even those caught in a web, but spiders of the family *Uloboridae* (approximately 300 species) have lost their fangs, forcing them to evolve their vice-like death shroud as their only way of directly killing other animals.[18]

The shape of **deer** or sheep's horns can tell you a lot about the animal's fighting style. Ungulates that sport horns with the tips facing downwards tend to ram each other, whereas those whose tips face inwards wrestle each other. Deer or antelope with smooth horns fight by trying to stab each other. The shape of the horns can also reveal something about how each animal lives. Those with horn tips facing in tend to be solitary, monogamous species whereas those with horns facing out tend to be polygamous animals that live in social groups.[19]

One species of **tortoise** (*Chersina angulata*) belies the sedate reputation of its brethren and is so aggressive that it has been nicknamed the fighting tortoise. Unlike many other species of tortoise, the *C. angulata* that lives in South Africa has a long gular, a horn-like protrusion that extends from the bottom shell. Males of the species use their gular when fighting to ram and overturn their competitors on to their backs, where they can be so helpless that they often spend over 20 minutes wriggling about without successfully righting themselves.[20]

There is a species of shrimp that will form a defensive army to repel animal intruders. When a nest of the **snapping shrimp** (*Synalpheus* spp.) is threatened by an intruder, a sentinel shrimp will recruit other colony members to snap their claws in concert for tens of seconds, creating a striking cacophony that is highly successful at repelling the invader. The shrimp do not rush to the site of the attack, but coordinated snapping appears to be a warning signal to would-be intruders that the nest is occupied by a cooperative colony ready to defend it. Such gang behaviour is a form of eusociality that is usually seen only in various species of mammals, birds and insects that form social communities.[21]

Some **bacteria** dupe others into doing some of the hard work of gathering food for them. Most *Pseudomonas aeruginosa* bacteria produce chemical structures known as siderophores, which they synthesise and excrete when they become deficient in iron. These siderophores bind any iron available in the environment, making it available for uptake by all the bacteria in the immediate vicinity. However, there is a mutant strain of *P. aeruginosa* that does not produce these binding chemicals and instead lets the others do all the hard work for them. The freeloaders then uptake the iron nutrients without putting in any of their own effort.[22]

One in 250 **dog** bites in the USA will fracture a bone of a person being bitten.[23]

In the UK alone, domestic **cats** kill 57 million mammals a year, 27 million birds and 5 million reptiles and amphibians.[24]

Domestic **cats** living in bungalows bring home greater numbers of birds than cats living in terraced houses or flats. Cats living in detached houses bring home greater numbers of mammals than cats living in semi-detached and terraced houses.[25]

Cats arch their backs when confronting each other as a way of demonstrating their size to competitors. It is thought that back-arching is a reliable signal of a cat's size as the skeleton, muscles and fur limits how big a cat can appear during the display.[26]

Swarms of **Mormon crickets** (*Anabrus simplex*), which in the USA can reach up to 10 kilometres long, keep on the move not just to find food, but to avoid being eaten by each other. If an individual cricket stops for any reason, it is likely it will be devoured alive by some of the millions of its cannibalistic brethren that are following.[27]

The parasitic **gordian worm** (*Paragordius tricuspidatus*) begins its life in water before infecting the bodies of a larger insect host: a cricket. But the worm has a remarkable and unique ability to survive even if its insect host is eaten by a larger predator, the only parasite known to do so. When the cricket is eaten, and partially digested, the worm escapes by burrowing through the body of the predator, usually a fish such as a perch, bass or trout, or an amphibian such as a frog, until it emerges unscathed. The worms have been seen emerging alive and well from the gills of predatory fish and the mouths of predatory frogs, before dropping back into the water where they continue their life cycle.[28]

The **European hedgehog** (*Erinaceus europaeus*) has immunity to the venom of a species of the South American **viper snake** (*Bothrops jararaca*), even though the two species do not live alongside one another. The hedgehog's muscles contain an antihaemorrhagic factor named erinacin that inhibits blood-thinning compounds found in the viper's venom.[29]

The accuracy with which sharks attack their prey can be judged from a survey made between 1990 and 2000 of endangered **Hawaiian monk seals** (*Monachus schauinslandi*) living around the north-western Hawaiian Islands. The seals are predated upon by the **tiger shark** (*Galeocerdo cuvier*) and the **Galapagos shark** (*Charcharhinus galapagensis*). Overall, 15.3 per cent of all shark bite injuries occur on the seals' snouts or shoulders, an area of the body least susceptible to fatal wounding. However, 25.6 per cent occur in the more vulnerable area between the shoulder and diaphragm and 33.8 per cent between the diaphragm and pelvic girdle, the main areas of the body where the most severe injuries occur, while 25.3 per cent of bites are made between the pelvic girdle and hind flipper.[30]

Carnivores with the strongest bites

(expressed as a bite force quotient adjusted for different body size, maximum prey size and feeding category)

Tasmanian devil (*Sarcophilus harrisii*)	181
Spotted-tailed quoll (*Dasyurus maculates*)	179
African hunting dog (*Lycaon pictus*)	142
Jaguar (*Panthera onca*)	137
Clouded leopard (*Neofelis nebulosa*)	137
Grey wolf (*Canis lupus lupus*)	136
Tiger (*Panthera tigris*)	127
Cheetah (*Acinonyx jubatus*)	119
Spotted hyena (*Crocuta crocuta*)	117
Brown hyena (*Hyaena hyaena*)	113
Lion (*Panthera leo*)	112
European badger (*Meles meles*)	109
Dingo (*Canis lupus dingo*)	108
Cougar (*Felis concolor*)	108
Singing dog (*Canis lupus hallstromi*)	100
Arctic fox (*Alopex lagopus*)	97
Leopard (*Panthera pardus*)	94

Red fox (*Vulpes vulpes*)	92
Coyote (*Canis latrans*)	88
Grey fox (*Urocyon cineroargentus*)	80
Brown bear (*Ursus arctos*)	78
Aardwolf (*Proteles cristatus*)	77
Jaguarundi (*Felis yagouaroundi*)	75
Black bear (*Ursus americanus*)	64
Domestic cat (*Felis sylvestris catus*)	58
Striped genet (*Gennetta tigrinus*)	48
Asiatic bear (*Ursus thibetanus*)	44

table source [31]

The terrifying carnivorous **dinosaurs** (*Allosaurus and Tyranno-saurus rex*) did not have a bite to match their reputation. Estimates suggest that both could bite down on prey with a force of 5 to 7 kilonewtons. A modern 3.7-metre-long alligator has been shown to exert bite forces up to 9.5 kilonewtons, meaning an alligator of roughly equivalent size to *Allosaurus and T. rex* would have a bite some ten times more powerful, being able to bite down with a force of 70 kilonewtons. [32]

Are marine or terrestrial carnivores more efficient hunters?

Hunting efficiency is measured as the ratio of energy acquired from the ingestion of prey compared to the energy expended during a single hunting event. For marine mammals a hunting event is defined as a single dive. A hunting event by terrestrial mammals is defined by the period to bring down a single prey item.

	HUNT DURATION IN MINUTES	KCAL EXPENDED	KCAL INGESTED	EFFICIENCY
Weddell seal (461 kg)	16.3	137	1,397	10.2
Sea otter (25 kg)	2.4	5	19	3.8
African lion (170 kg)	180	5,062	19,498	3.8
Wild dog (25 kg)	104	1,288	2,836	2.2

table source [33]

Beaver (*Castor fiber*) is considered a delicacy in some parts of the world. On average, a beaver carcass contains 63 per cent meat, 14 per cent fat and 23 per cent fat. It will yield 5.5 kilograms of meat, two-thirds of which comes from the thighs. In order of meatiness, the most valuable parts of a beaver carcass are the thigh, flank, loin, shoulder and tail.[34]

The average person in the UK will eat 550 poultry, 36 pigs, 36 sheep and 8 cattle in their lifetime.[35]

A plucked **goose** carcass has a density of 937 kilograms per cubic metre.[36]

Biomechanical wonders

❧

THERE IS JUST ONE invertebrate known that is capable of getting an erection. Erect penises and clitorises are common in mammals but erectile tissue is virtually unknown among the soft-bodied invertebrates. However, the male **octopus** (*Octopus bimaculoides*) engorges its sexual organ known as the ligula – which also doubles as the tip of one of its eight arms – with blood, stiffening it in order to inseminate a female.[1]

Despite having bendy arms that can move and bend anywhere along their length, octopuses use a distinctly human method of bringing food to their mouths. Human arms are restricted by having fixed joints, which allow us to bring food up to our mouths by rotating our arm around the shoulder, elbow and wrist. An **octopus** (*O. vulgaris*) uses a similar strategy to feed itself, bending its arms at three particular discrete points that mimic the three joints in the human arm. Its arms also stiffen around these joints, much like the stiff bones in a human arm. Bringing food up to its mouth using this technique appears to be the most efficient way for an octopus to pick up and eat food.[2]

For over a century, biologists have wondered how **chameleons** use their amazing tongues to catch prey, such as insects, reptiles and even birds, from up to two body lengths away. The answer is twofold. Chameleons are able to shoot out their tongues at lightning speed by storing up energy in the organ, rather like energy is stored within a catapult that is pulled taut. Around 200 milliseconds before a strike, chameleons use their tongue muscles to store energy in parts of the tongue called the intralingual sheaths, which pack into each other like the collapsing parts of a telescope. As they strike, this energy is released within 20 milliseconds, accelerating the tongue pad forward.[3]

Chameleons then use another trick to reel their prey back in. The muscles of most animals are not very strong when fully extended, as can be understood by trying to lift a heavy weight with an outstretched arm. But chameleons have a unique set of tongue muscles that can exert maximal force even when extended, providing enough strength to pull in distant prey.[4]

Frogs stick out their tongues in three different ways. Primitive species of frogs protract their tongues using a mechanism called mechanical pulling. These animals contract a muscle called the *m. genioglossus*, which causes the tongue to bunch at the front of the jaws and extend beyond. But the tongues of such frogs are so short – less than 70 per cent of the length of their jaw – that they cannot get near prey without lunging their whole bodies forward.

Other **frogs** use a technique called inertial elongation, which has evolved at least seven times in anurans. Tightly coordinated tongue and jaw movements flip the tongue over the mandibles and the tongue shoots out way beyond its resting length. This happens extremely fast, and also requires little movement on behalf of the frog, which can stay hidden while shooting for prey.

But this is a ballistic shot and the frog cannot change the direction of its tongue mid flick. The tongue also shoots out in whatever direction the frog's head is pointing.

Then there is the ultimate **frog** tongue, powered by a technique called hydrostatic elongation. This mechanism is similar to inertial elongation, but the tongue has a separate compartment of dorsoventrally arranged muscle fibres that are surrounded by connective tissue. As these muscle fibres shorten, out shoots the tongue. The difference however, is that the tongue can be aimed both left and right, and up and down relative to the head. Microhylid frogs that have hydrostatic tongues can catch prey sitting over 90 degrees to the side of their head, without moving their bodies.[5]

Animals usually swallow using their tongue and throat. Frogs use their eyeballs. When the **northern leopard frog** (*Rana pipiens*) swallows food such as a small cricket, it closes its eyes and retracts its eyeballs into its body. These push into the pharynx and against the prey item, and regular retractions help force the food to the back of the oesophagus.[6]

Most frogs love water, but arboreal tree frogs, which can live in arid environments, work hard to waterproof their bodies. The **South American frog** (*Phyllomedusa sauvagei*) and the **Australian hylid tree frog** (*Litoria caerulea*) secrete lipids from specialised glands within their skin then perform a complex series of self-wiping movements with their limbs that spreads the lipids over their bodies. The lipids create a waterproof layer that helps prevent moisture evaporating from the frogs' bodies.[7]

Some birds jump like frogs. Tiny tree frogs use a biomechanical trick to power their leaps, storing elastic energy in their tendons which is released when they jump. That gives a greater power boost than they could achieve using their muscles alone. **Guinea fowl** (*Numida meleagris*) use a similar technique. A guinea fowl's leg muscles can produce only 330 watts per kilogram of power. But when the birds leap into the air, they produce nearly 800 watts per kilogram of power, with the rest being provided by energy stored in tendons in their legs.[8]

It may not appear so to the casual observer but most species of **birds** are able to move their upper bill as well as their lower bill. In a process called prokinesis, the birds have a hinge-like mechanism between their upper bill and skull that allows the upper bill to bend.[9]

A spinning **hen's** egg can literally jump off the table. If a hard-boiled egg is spun at a speed of 1,500 rpm it will spontaneously leave the table, making a series of tiny leaps each just a fraction of a millimetre high and lasting for just 0.01 second. Apart from the spinning motion, no other external force is required. It is not yet known whether a raw egg also leaps this way.[10]

Ostrich shells have been used as water vessels to carry fresh water for millennia. Namibian aborigines have preserved this knowledge and profit from it by using ostrich shells to ferry water during their long desert migrations. But it is not just the size of the ostrich shell that makes it such a suitable water container. The shells of **ostriches** (*Struthio camelus*) also have antibacterial and antifouling properties.[11]

The mature oocyte, or yolk, within an ostrich's egg is the largest single cell found in nature.[12]

The aptly named **archer fish** (*Toxotes jaculatrix*) has an impressive ability to shoot down its insect prey. The fish shoots a precisely aimed jet of water out of its mouth to knock off insects resting on leaves or stems overhanging the water. What is more, within 100 milliseconds of the prey being knocked off its perch, the fish will turn and begin swimming to where its victim will land in the water. The fish can predict where the insect will land so precisely that it arrives at the point of prey splashdown just 50 milliseconds after the insect has hit the water, ensuring it is first to the meal.[13]

Peacock mantis shrimp (*Odontodactylus scyllarus*) are champion underwater boxers, wielding a pair of specialised club-shaped limbs that they use to smash snail shells into smithereens. But like human boxers, the shrimp, which are the fastest strikers in the animal kingdom, are not content landing a single blow. Instead, each strike of their powerful biological hammers rains two successive blows on a shrimp's hapless prey. The first blow comes simply from the force of the hammer landing. But each strike does more than that, as it also creates an area of low pressure as the hammer rebounds from the snail's shell. This causes vapour bubbles to form, which implode with such intensity that they generate a second impact that lands just 390 to 480 microseconds after the first. With each strike, the tiny shrimps can land an impact force equal to thousands of times their body weight.[14]

Some sponges have skeletons made of glass. Like many sponges, the aptly named **glass sponge** (*Euplectella* spp.), a deep-sea, sediment-dwelling sponge from the western Pacific, has a skeleton made of structures known as spicules. In many sponge species, these are made of an organic material called spongin, but in the glass sponge, as well as other species, they are made of silica. Not only do these glass spicules form a strong structure, being more flexible and less brittle than equivalent synthetic glass rods of the same length, but they also have optical properties similar to man-made optical fibres. The whole skeleton of the glass sponge is essentially one large cylindrical glass cage that has seven different levels of structural organisation. The resultant structure might be regarded as a textbook example of mechanical engineering, because the seven hierarchical levels in the sponge skeleton represent major fundamental construction strategies such as laminated structures, fibre-reinforced composites and bundled beams.[15]

The self-sharpening teeth of **sea urchins** are designed in a similar way to fibre-reinforced plastic, with fibres running through a surrounding material. Sea urchin teeth have calcite fibres interweaved within a wider body of calcite comprising crystals of different shapes and sizes. Levels of magnesium increase towards the middle of each tooth making the tooth significantly harder on the inside than the outside. This in turns means that as the tooth wears, it progressively reveals ever harder material, allowing each tooth to stay sharp naturally.[16]

The manufacturer of nature's strongest glue is a **bacterium** (*Caulobacter crescentus*), a bug that lives almost anywhere wet, and is notoriously difficult to shift from the surfaces upon which it chooses to make its home. This is because it attaches itself to a substrate by making a sticky substance out of a string of long sugar-based molecules called polysaccharides. The average *C. crescentus* requires around 70 newtons per square millimetre to rip it from a surface, a grip that is around seven times stronger than that produced by a gecko's foot, and three times stronger than commercial superglue that gives up at about 25 newtons per square millimetre.[17]

The long-necked **sauropod dinosaur** *Diplodocus*, whose neck stretched for some 9 metres, could only browse on grasses and low-lying trees rather than browsing on the upper branches. For although its neck was long, it was not flexible, and *Diplodocus*'s skeleton and vertebrae did not allow it to raise its head very high.[18]

Large **sauropod dinosaurs** such as *Brachiosaurus* would not have been able to swim. They were so buoyant that most of the giant reptile's body would have floated above the water and the slightest waves would capsize the animal. The dinosaurs would have been extremely unstable even in shallow depths.[19]

For most **mammals**, the mass of their skeleton is directly proportional to the mass of their whole body.[20]

The giraffe, manatee and sloth all have one unique feature in common, the unusual number of vertebrae they have in their necks. All mammals, from mice to whales, with the exceptions above, have seven cervical vertebrae, the bones in the spinal column that begin above the thoracic vertebrae to which ribs are attached. **Giraffes** (*Giraffa camelopardalis*), however, have eight cervical vertebrae, whereas **manatees** (*Trichechus* spp.) have six. **Two-toed sloths** (*Choloepus* spp.) have a variable number of either six or seven cervical vertebrae, whereas **three-toed sloths** (*Bradypus* spp.) have nine.[21]

The appropriately named **long-tailed pangolin** (*Manis tetradactyla*) holds the record for having more vertebrae in its tail than any other mammal, possessing 47 in total. The pangolin, which looks like an anteater covered in thick plates made of keratin, has a prehensile tail, which it uses to climb trees and hang from branches in the search for arboreal ants or termites.[22]

Snakes have super-stretchy skin. The skin of the **garter snake** (*Thamnophis sirtalis*), for instance, stretches considerably around the circumference of the snake's body, an adaptation that allows the snake to swallow large animals and fit them inside its own body. Skin covering the snake's body from the mouth to the stomach, which has to accommodate whole prey, is more stretchy than skin covering the lower parts of the animal where sections of the digestive tract have to cope only with food that has already been partially digested.[23]

The shake of a rattlesnake's tail is one of the fastest movements of any vertebrate. **Rattlesnakes** can shake their rattles at a frequency of 100 hertz for hours.[24]

A **western diamond-backed rattlesnake** (*Crotalus atrox*) rattles its tail at a progressively higher frequency as it matures from a juvenile to an adult, increasing the rate at which it rattles 20-fold as the snake grows from around 20 grams to 100 grams in bodyweight. But then old age sets in, and an older rattlesnake is no longer able to rattle as fast as it once could. As the snake matures it can increase its rattle speed because the contractile muscles it uses to shake its tail become more aerobically efficient. But once a snake reaches a certain age, and grows to around 1 kilogram in weight, it develops a more ossified, and thus disproportionately heavier, bone in the base of the rattle. That makes it harder for a large rattlesnake to move its rattle to and fro, and the contraction frequency of its tail decreases by around 30 per cent.[25]

Sperm vary more dramatically in size and design across animal species than any other type of cell.[26]

A cat's ability always to fall on its feet is due to a perfect body design. A falling **cat** uses its eyes, and part of the inner ear called the vestibular apparatus, which monitors balance, to work out which way is up and down. Once the cat knows where the ground is, it uses its neck muscles to turn its head to an upright horizontal position, and then rapidly aligns the rest of the body the same way.[27]

There is a bird that routinely exercises by doing push-ups with its wings. Nestlings of the **common swift** (*Apus apus*) perform the trick because they cannot be sure how much food their parents will bring back to the nest or how heavy they have become. This creates a very real danger as, if they are too heavy when they try to take to the sky, they will come crashing down to land or fly too slowly to avoid predators. And once a swift has taken to the air, there is no turning back because fledglings do not return to their nests and will not land again until future breeding seasons when they themselves have grown old enough to reproduce. So, to ready themselves for flight, the young birds do push-ups using the exercise as a way to gauge their strength and weight. By doing this, and flapping their wings in the nest, the nestlings are able to estimate their wing loading accurately. Before they fledge, most swift nestlings adjust their body weight to match the loading of their wings and ensure they can take to the sky in peak condition.[28]

Marsupials have larger hearts on average than placental mammals.[29]

Black bears (*Ursus americanus*) have an amazing ability to stop their bones degenerating, even as they get older or during hibernation when they lie idle for five to seven months on end. In most animals and people, bones get weaker if they are not used.[30]

When **dinosaurs** known as *Therizinosauroids* emerged from their eggs, their legs were already strong enough to allow them to run after and chase down prey. They also already possessed teeth capable of ripping through meat.[31]

There is a spider that literally uses a rod, line and bait to angle for its prey. **Bola spiders** spin a short fishing line of silk. At the free end they attach a sticky globule, or bola, which releases scent chemicals called pheromones that mimic the scent of certain female moths. The spider holds the fishing line with one leg and hurls the bait at inquisitive male moths that may fly close by. The male moths are attracted to the bait, expecting it to be a female moth, and get stuck. The spider then reels in its catch and settles down to dinner.[32]

Some spiders do not have to bite their prey to inject them with venom. Instead they spit poisonous silk at their unsuspecting victims, which both snares and paralyses the prey. These **spiders**, called the *Scytodes* species, have venom glands that are connected to the usual silk glands located at their rear. From these silk glands these tiny spiders can shoot their poison silk relatively large distances.[33]

The webs of **orb web spiders** (*Araneus cavaticus*) are not actually sticky to begin with. When orb web spiders weave their wonderful creations, they first lay down a non-sticky scaffolding of silk. One part of this scaffolding, the radii, is arranged like the spokes of a wheel. The spider then subsequently attaches another sticky spiral (or adhesive or viscid spiral, among other names) to the first radii. Between 40 and 70 per cent of an orb web spider's web is not made from silk. Although silk provides the structure of the web, much of its mass actually comes from organic and inorganic chemicals known as low-molecular-mass compounds, such as GABamide and glycine, phosphorylated glycoproteins and lipids that help form the web's sticky adhesive coating.[34]

A spider's web is not equally strong throughout its structure. **Spiders** spin silk of constantly varying tensile properties and a single web created by a single spider will contain a wide variation of stronger and weaker silk threads.[35]

Parrots produce their gloriously coloured feathers in a way like no other bird. Other birds use pigments known as carotenoids that are found in food to produce colours such red or yellow. But parrots, such as the **eclectus parrot** (*Eclectus roratus*), do something different. Their feathers are coloured red by an unusual set of pigments called psittacofulvins that the birds synthesise within their bodies, and the higher the concentration of psittacofulvins the redder the bird. No other animal is known to produce such colourants.[36]

There are **birds** in New Guinea that have poisonous feathers and skin. The birds, five species that belong to the genus *Pitohui*, eat *Choresine* beetles, and in doing so digest neurotoxic chemicals called batrachotoxins that are produced by the insects.[37]

The antlers of **red deer** (*Cervus elaphus*) are one of the hardest biological materials, being as strong as bone or ivory. But they can be softened, worked into different shapes and then hardened again. The process is achieved by soaking the antler in water which makes it malleable and flexible. Air drying the antler then returns it to its original hardness.[38]

According to ancient myth, **hippos** sweat blood. Not true, as the animals lack sweat glands. And while water-loving hippos do secrete a clear liquid all over their body that turns reddish brown, it is actually a multipurpose sunscreen, coolant, and antibiotic all in one. The secretion contains two chemicals, hipposudoric acid (a red pigment) and norhipposudoric acid (an orange pigment), both of which block visible and ultraviolet light. This hippo 'sweat' is thought to help protect against damaging light rays and also help cool the animal. The red pigment also works as an antibiotic, which could help clean the wounds of fighting males.[39]

The **Galapagos marine iguana** (*Amblyrhynchus cristatus*), the only sea-going lizard, is also the only vertebrate known to shrink in body size regularly when adult, and then grow larger again. The iguanas shrink up to 15 per cent in body length during El Niño weather events, losing bone mass as an adaptation to survive more harsh weather conditions. The following year the iguanas grow even larger than they were before shrinking.[40]

Marine **iguanas** living on different islands also vary in size by an order of magnitude. Iguanas living on Genovesa Island in the north-east of the Galapagos archipelago grow to a maximum body weight of 1 kilogram, while the largest iguanas on Isabela Island in the south-west of the archipelago can weigh 12 times heavier, reaching weights of over 12 kilograms.[41]

Super organs

❧

Tiger snakes can happily do without their eyes. **Snakes** (*Notechis scutatus*) whose eyes have been pecked out by seagulls grow and mate just as well as their seeing counterparts, the first direct evidence that some animals still possess organs and abilities they can do without.[1]

Tarsiers, a primitive group of primates, have eyeballs that are bigger than their brains.[2]

Jumping spiders have unique eyesight. Their complex eyes give them a spatial resolution ability superior to any other animal of their size.[3]

Box jellyfish (*Tripedalia cystophora*) have four types of eye: two smaller types of eye that contain pigment and are shaped as either slits or pits, and two either large or small types of eye containing lenses. Each type of eye records different information such as light levels. The jellyfish is thought to have four types of eye because it is too simple an organism to process more information through a single, more complex organ. These jellyfish also have four parallel brains and 64 anuses.[4]

Lizards have three eyes. As well as the two obvious eyes at the front of the head, they have a third parietal eye on the top of the head. The eye has a rudimentary retina and lens but it cannot form images. Instead it is sensitive to blue and green light, and lizards are believed to use it to detect the arrival of dawn and dusk.[5]

Domestic cats (*Felis sylvestris domestica*) have eyes containing slit pupils, while **Siberian tigers** (*Panthera tigris altaica*) have eyes with circular pupils. In dogs, the **European red fox** (*Vulpes vulpes*) has slit pupils, while **grey wolves** (*Canis lupus lupus*) and **domestic dogs** (*Canis lupus familiaris*) have round pupils.[6]

Parrots have much larger brains relative to their body size than any other group of birds, which may explain their unique abilities among bird species to perform complex mental tasks and speak human-like words. A number of parrot species even have relatively larger brains than comparably sized primates.[7]

Many species of **bat** have evolved to have smaller brains than their recent ancestors.[8]

Carnivores that live at high latitudes where there is lots of snow tend to have longer penises.[9]

Males of **bat** species that have larger brains also tend to have smaller testicles. What is more, male bats that have to compete for females, which like to mate with many males, tend to be pea-brained compared to species in which females remain faithful to one partner.

Bats display an extraordinary variation in the sizes of their testes, with some species having tiny testicles that constitute just 0.12

per cent of their body mass, whereas others have huge gonads that account for 8.4 per cent of their body mass. That is a greater range of testicular size variation than in any other group of mammals. For example, different primate species have testes that vary from 0.02 per cent of their body mass to 0.75 per cent.

The reason that pea-brained bats have larger testicles is because of the energy required to grow each organ. Bats that put a lot of energy into growing large testes, required to produce enough sperm to compete for and inseminate many females, have little metabolic energy left to develop their brains. Whereas males in monogamous relationships can put less energy into having sex, and more into developing their mental abilities.[10]

Most **mammals** have 100,000 neurons underneath every square millimetre of the surface of their brain cortex. Primates have up to 200,000 neurons under each square millimetre.[11]

An **alligator's** (*Alligator mississippiensis*) tail accounts for 28 per cent of its total body weight.[12]

The scaly ant-eating **giant pangolin** (*Manis gigantean*) has a tongue as long as its tail, both of which measure around 70 centimetres from end to end.[13]

An average **ram's** scrotum increases in circumference from 15 centimetres when the ram is 12 weeks old to 25 centimetres when it reaches 35 weeks old.[14]

A deer's antlers are traditionally thought to serve no other purpose than simply being weapons with which to fight other males. Large antlers are good weapons as well as signalling to other males that their bearer is likely to be a strong fighter. Also the fact that males with large antlers father more calves is often thought to be because these males win more fights and form larger harems.

However, the size of a **red deer's** (*Cervus elaphus hispanicus*) antlers also directly indicates how large a stag's testes are, how many sperm a stag will produce and how fast those sperm will swim. Therefore a stag with large antlers is not only more likely to win more fights and have access to the most females but he is also more likely to impregnate a female when he does mate. Knowing this, female deer are also more likely to be attracted to stags with large antlers.[15]

The fleshy tentacles of the **star-nosed mole** (*Condyhru cristutu*) have evolved differently from the appendages of any other animal. The legs of mammals, insects' antennae or wings, or the appendages of other invertebrates all evolved as outgrowths of the body wall controlled by a similar group of genes. But the mole's tentacled nose evolved from swelling on the side of the face that expanded and then folded outwards.[16]

Aquatic **newts** are the only animals known that can regenerate the lens in their eyes if it becomes damaged or is removed. The newts, a type of urodele amphibian, have the most extensive regenerative abilities of any known animal, also being able to grow new limbs such as legs and tail.[17]

Male birds that have big brains also have healthy immune systems. Those species with a large amount of brain matter relative to their body size, such as **yellowhammers** (*Emberiza citrinella*) and **barn swallows** (*Hirundo rustica*), also have relatively larger spleens and bursa of Fabricius, two organs crucial to the immune system. A large brain may be needed so males can sing elaborate songs, and in turn advertise their health to females.[18]

Laboratory mice have been studied in detail for decades, but it was only in 2006 that it was discovered that the rodents possess a second thymus gland in the neck, a vital organ for producing immune cells required to fight off infection.[19]

Centipedes' bodies are arranged into a series of segments, each sporting a pair of legs. Different species can have anything from 15 to 191 individual segments, and different centipedes of the same species may even have a different number of segments to each other. Bizarrely, though, the number is always an odd one. No **centipede** is known to have an even number of body segments.[20]

Genetically engineered **mice** have been developed that have hearts that glow green every time they beat.[21]

Energy efficient

❧

THE **Madagascan fat-tailed dwarf lemur** (*Cheirogaleus medius*) is the only primate in the world that hibernates. It is also the only animal living in the tropics known to do so. Most animals hibernate to conserve energy during harsh cold winters, but the fat-tailed dwarf lemur hibernates in a tree hole even when the temperature soars over 30°C. It is thought to hibernate to avoid the huge swings in temperature that bedevil the tropics during winter days.[1]

Woodchucks, marmots and squirrels reduce their metabolic rate more quickly going into hibernation than they do when undergoing hypothermia.[2]

The **hoary bat** (*Lasiurus cinereus*) is unique among free-ranging mammals in that it hibernates for just a few days at a time. What is more, the hibernating is done by pregnant females, which go into a deep state of torpor to tough out harsh weather conditions, ensuring they have enough energy left to feed milk to their offspring.[3]

After not eating for just ten days, **polar bears** (*Ursus maritimus*) can enter a 'fasting state' during which their bodies shut down to consume energy. Biochemically and physiologically their bodies undergo similar changes to that seen during hibernation, except that it can happen at any time of the year and the bears are still capable of hunting, with their bodies reverting to their normal state as soon as a meal is consumed. Polar bears can lose up to 1 kilogram of weight each day when searching for food.[4]

Most reptiles keep cool by opening their mouths. But one, the **Gila monster** (*Heloderma suspectum*), keeps cool using an unusual orifice – its reptile equivalent of a back-passage. The venomous lizard, an active predator that lives in the desert where temperatures rise well above 40°C, keeps its body temperature 3°C below the ambient temperature using 'cloacal cooling', shedding water from its bladder to the atmosphere out of its rear opening called a cloaca. No one knows why the lizard cools itself this way, but one suggestion is that opening its mouth would compromise the quality of its venom, which is released straight into the mouth like saliva.[5]

Female **reindeer** eat more and get fat during the winter. But contrary to expectation, these extra fat reserves do not help the reindeer give birth to more offspring in the spring, produce more milk or heavier calves, or give either the mothers or offspring any greater chance of survival. Instead, it is thought the reindeers slim down again by the time spring arrives, because leaner, meaner mothers have a better chance of defending their young against predators.[6]

The metabolic rate of a walking **ant** (*Camponotus* spp.) is fourfold higher than that of a resting ant. An ant walking up a 30-degree incline has a sevenfold higher metabolic rate than a stationary one.[7]

A **gecko** (*Phelsuma dubia*) can run faster up a wall when the temperature is around 25°C than when the temperature is a milder 15°C or warmer 35°C.[8]

A camel will increase its body temperature during the day but decrease it at night. The adaptation helps the **camel** (*Camelus dromedarius*) maintain its body temperature during the harsh desert environment and prevents the camel losing precious water reserves by sweating or panting to keep cool. The **Arabian oryx** (*Oryx leucoryx*) and **Arabian sand gazelle** (*Gazella subgutturosa marica*) are thought to use the same trick.[9]

During slow, level flight the pectoral muscle of a **pigeon** (*Columba livia*) performs 24.7 joules per kilogram of body weight of work.[10]

When jumping, each hind leg of a **horse** (*Equus caballus*) performs 0.71 joules per kilogram of body weight of work. The knee does 85 per cent of the work (0.60 joules per kilogram of body weight) of each limb.[11]

A domestic **goat** expends between 3.35 and 3.63 joules per kilogram of body weight walking in a straight line along a flat surface.[12]

Aestivating frogs and toads, which undergo a similar process to hibernation, shrink their intestines in order to survive fasting for many months before they eat their next meal. Three distantly related species of **frog** and **toad** (*Bufo alvarius*, *Ceratophrys ornata* and *Pyxicephalus adspersus*) shrink the mass of their small intestine to between one-half and one-third of its normal size. When they do finally snack on prey, their intestines increase in size again and their total capacity to uptake nutrients through their intestines rises six to tenfold.[13]

Depending on the species, a female **bird** exerts between 16 and 27 per cent more energy when producing an egg, as measured by the rise in her resting metabolic rate.[14]

Penguins have an uncanny ability to know how high they should leap out of the water to land on sea ice. **Adélie penguins** (*Pygoscelis adeliae*) look to see the reflected image of the height of ice above the water and adjust the angle at which they leap from the water accordingly. **Emperor penguins** (*Aptenodytes forsteri*) are also able gauge the height of the ice surrounding an ice hole, and they use this information to adjust the speed at which they burst out the water, ensuring they clear the edge of the hole and land on the ice, however high it is.[15]

Swordfish heat up their eyeballs when they are going hunting. The trick allows them to see better and pick out the moving shape of fleeing prey. Although most fish are cold-blooded, swordfish are able to heat their eyes to 28°C, even in nearly freezing water. At this temperature they can spot a moving squid up to 12 times faster than if they were seeing with cold eyes. It may also allow the swordfish to see the squid before the squid sees the swordfish.[16]

Hummingbirds shut down their kidneys at night. Because of their incredibly small size and very high rate of metabolism, **broad-tailed hummingbirds** (*Selasphorus platycercus*) face an unusual dilemma of keeping just the right amount of water in their bodies.

During the day, hummingbirds ingest extraordinary amounts of water from the nectar upon which they feed, most of which must be excreted. But during the night, or when they spend long periods flying, they face the opposite problem, one of conserving enough water to keep their bodies going. Slowing their kidneys down prevents too much water being lost.[17]

Adélie penguin (*Pygoscelis adeliae*) chicks take part in group-hugs of up to 18 individuals in a bid to stop them freezing to death. Smaller chicks appreciate a cuddle the most as they lose heat much faster than their larger brethren.[18]

The **wood frog** (*Rana sylvatica*) freezes solid during the winter before thawing out as the temperature rises in spring. The frog has a unique physiology that prevents damaging ice crystals forming within its cells.[19]

Ants tailor the size of their houses to how many residents live within them. Despite there being thousands of **worker ants** (*Messor sancta*) within any one colony, somehow the insects always coordinate their activities to build a perfectly sized nest, adjusting the volume of cavities and tunnels according to how many workers need a home.[20]

Rattlesnakes (*Crotalus durissus*) raise their body temperature to help them digest their latest meal. The larger the meal, the hotter their bodies become.[21]

Female birds can enlarge and shrink their reproductive organs within days. When not in use, the organs shrink down, as they are metabolically expensive to maintain. But as soon as the birds are ready to mate, lay eggs and reproduce, they grow a large ovary and oviduct. In **starlings** (*Sturnus vulgaris*), females increase the size of their reproductive organs 22-fold.[22]

Small **birds** will lose up to 5 to 10 per cent of their body weight overnight.[23]

Birds are able to take off faster and at steeper angles after migrating than before, due to the fact that they can shed between one-third to two-thirds of their body weight during their travels.[24]

The average male **Japanese macaque** (*Macaca fuscata*) ejaculate has an energy, or calorific, content of 8.1 kilojoules. Depending on the individual's body mass and the number and volume of the ejaculates, macaque males are thought to use between 0.8 per cent and 6.0 per cent of their energy per day producing ejaculate during the breeding season.[25]

For their weight, large **mammals** use a lot less energy to run a given distance than small mammals do.[26]

Life is easy for one particular group of mole rats. Like their **naked mole rat** (*Heterocephalus glaber*) cousins, **Damaraland mole rats** (*Cryptomys damarensis*) live in social colonies. Together the two species are the only known eusocial mammals, with the basic idea of eusociality being that closely related individuals live and work together to ensure the success of the colony. However, some Damaraland mole rats make a mockery of this idea. In their society, work is divided between industrious mole rats that do over 95 per cent of all the work, and lazy mole rats that do less than 5 per cent. The lazy mole rats constitute a physiologically distinct caste that apparently refuse to work and help the queen reproduce. Instead they tend to sit about, eat and get fatter than their more work-orientated relatives. The only time the lazy workers get off their mole rat backsides is when it rains and the soil becomes loose. Then they pitch in, helping the colony to dig new territories and disperse.[27]

Among dolphins, all blubber is not equal. Over the first three months of a young **Atlantic bottlenose dolphin's** (*Tursiops truncatus*) life, as it develops from a tiny baby to a plump juvenile, a dolphin will grow thicker blubber that contains more fat, leading to a threefold increase in its insulating properties. Pregnant females and young adults tend to have blubber that has the highest fat content, but as dolphins age, the fat content of their blubber reduces. The blubber composition of two individual dolphins of the same species is also often radically different, as one adult may have a thick layer of blubber with a fat content of just over 3 per cent, while another might have blubber with a fat content of over 40 per cent.[28]

Do the locomotion

A GOLDEN RULE DESCRIBES how often **fish** beat their tails, **birds** flap their wings and even why bats cruise through the sky at the speed they do. This rule of thumb is called a Strouhal number, which is a measure of how efficiently an animal moves. The number describes how much up or down movement a wing or a tail makes relative to its owner's forward speed, calculated as stroke speed multiplied by size, divided by forward speed. And all animals seem to move at a Strouhal number of between 0.2 and 0.4.

The reason why all animals seem to move according to this rule is because wings and tails create eddies in the surrounding air or water. These eddies create turbulence and slow down the bird or fish moving through them. If they flap their wings or fins too quickly, they end up moving through eddies created by the previous stroke. Flap them too slowly and the turbulence sticks, slowing the animal further.

But somewhere in between is the perfect rate of flapping for each animal's size. And everything from blue whales, to mackerel, locusts, pigeons and bats all obey the same rule. Only very small insects seem to do things differently, perhaps because they are so tiny that the air around their bodies behaves differently, acting more like thick treacle than a flowing fluid capable of creating eddies.[1]

A **greyhound's** finely honed running style allows it to run around a tight and steeply banked racing track without slowing down. When a person runs around a banked bend, say on an indoor running track, their body experiences gravitational and centrifugal forces that effectively make their bodies heavier. To ensure their legs do not become overloaded by such forces, a human sprinter places each foot on the ground for a longer period of time than they would if they were sprinting along a straight track. And that slows them down.

Greyhounds also experience the same forces when running around a tight curve, but they do not run any slower and do not change the length of time each paw spends on the ground. This is because, compared to humans, greyhounds run using more muscles in their hips, can extend their spines, and place much of their weight on their front legs. This allows them to increase the load on their legs by 65 per cent without slowing.[2]

Postal **horses** have been used to deliver express mail for two millennia, spanning a period lasting from 540 BC to AD 1861. Each of these systems was remarkably consistent in the speed at which they ran their horses and the distances between horse-changing stations. Both were chosen to optimise the efficiency of the horses, by ensuring they could deliver the mail as quickly as possible without become too tired or falling lame. The most recent and well-documented horse-based mail system is the Overland Pony Express, which operated during 1860–61 in the USA between Sacramento, California, and St Joseph, Missouri, for the benefit of west-coast pioneers. This, and the mail delivery systems that preceded it, ran each horse for a distance of 20 to 25 kilometres at an average speed of about 16 kilometres an hour.[3]

Horses have an inbuilt turbocharge system that allows them to run especially fast over short and middle distances. When running intensively at high speeds, a horse's spleen will dump huge numbers of red blood cells into the bloodstream, raising haematocrit levels from a usual 40 per cent to over 65 per cent, which in turn increases the amount of oxygen that can be delivered to the horse's muscles. Over long-distance endurance races, the horse is limited by its ability to sweat and lose heat, and its haematocrit level remains at between 46 and 49 per cent.[4]

When walking, most **mammals** use what is called a lateral-sequence gait, where the placing of a hind limb on the ground is followed by the placing of the front limb on the same side of the body. However, primates use what is known as a diagonal-sequence gait; after one hind limb makes ground contact, the forelimb on the opposite side is the next to touch down.[5]

Animals evolved legs not to walk on land, but to walk under-water. A number of species of modern fish, such as bottom-dwelling antennariid **anglerfishes**, have fleshy limb-like pectoral and pelvic fins that they use to walk about. They even have two distinct styles of walking that are the same as those used by land-dwelling tetrapods, suggesting that underwater walking came first. Recent fossil fish from the Upper Devonian period, some 400 million years ago, such as *Acanthostega*, have fully developed tetra-pod limbs despite being gill-breathing underwater animals.[6]

A standing or trotting **horse** puts 57 per cent of its weight on its front legs and 43 per cent on its back legs.[7]

Dogs place 63 per cent of their weight on their forelegs and 37 per cent on their back legs.[8]

Kangeroos (*Macropoidea*) walk using five limbs. At slow speeds they use their tail as an extra limb, becoming pentapeds, rather than quadrupeds.[9]

It may not appear so, but kangaroos and wallabies both hop and jump, and the two require vastly different amounts of energy.

When hopping, a **yellow-footed rock wallaby's** (*Petrogale xanthopus*) muscles use an average 155 watts per kilogram of power. When jumping, say onto a rocky ledge, the animal changes the angle of its legs, and invokes the power in muscles within its legs, back, pelvis and tail, using up to 500 watts per kilogram of power.[10]

There are two types of octopus that can walk bipedally, using just two of their eight appendages to walk along the sea floor. This ability to walk on two legs was thought to be restricted to creatures that possess muscles attached to bones or other skeletal structures. However, like other octopuses, the **coconut octopus** (*Octopus marginatus*) and the **algae octopus** (*Octopus aculeatus*) are invertebrates that do not have skeletons, instead relying on flexible muscles supported by the fluid inside them. Both species are thought to wrap their other six arms around their body as they walk, disguising themselves as either coconuts or blobs of algae floating around underwater.[11]

When a **cockroach** (*Periplaneta americana*) is standing still, it moves both its antennae randomly. But when it begins walking, the insect automatically synchronises the motion of both antennae, moving them in a coordinated pattern.[12]

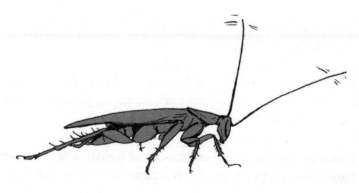

The **American cockroach** (*Periplaneta americana*) can execute as many as 25 individual turns per second when running along a wall.[13]

A waddling **penguin** uses up to twice as much metabolic energy when walking as other birds of the same mass. However, some turtles use just half as much metabolic energy walking on land as other terrestrial animals of a similar mass.[14]

Every goldfish swims in a unique way. Although all **goldfish** (*Carassius auratus*) look alike and are similarly built, they all swim with a particularly individual style and according to a unique pattern. Indeed it is possible to identify individual goldfish simply from the way they swim.[15]

Humpback whales (*Megaptera novaeangliae*) are outstandingly agile swimmers, renowned for throwing themselves out of the water and undertaking acrobatic underwater manoeuvres to catch prey. And their secret is having bumpy flippers. Called tubercules, the bumps increase the hydrodynamic performance of the whale by delaying stall, increasing lift and decreasing drag in the water.[16]

Spinner dolphins (*Stenella longirostris*), which make spectacular twisting leaps out of the water, are not actually able to twist their bodies in mid-air. They are hindered by the design of their skeleton, and instead rotate the body in a corkscrew fashion around its longitudinal axis underwater before breaching. Their angular momentum then takes over, spinning the dolphin around and around during the leap. However, the dolphins are still agile enough to spend up to 1.25 seconds airborne on each leap.[17]

A **great white shark** (*Carcharodon carcharias*) can swim at an average sustained speed of 4.7 kilometres per hour over a distance of 20,000 kilometres.[18]

Dogs have an amazing ability to retrace their steps. If they have their eyes and ears covered up and are then taken for a long walk, dogs can instantly find their way back to their exact starting point.[19]

The distance a **frog** can jump is not limited by the power of its back legs, but by the strength of its front legs, which have to take the brunt of the force of landing.[20]

Frogs also generate more force in their legs while jumping than swimming, even though both involve extending the legs backwards as the frog pushes against either the ground or water. On average, a **green frog** (*Rana esculenta*) will produce about twice the impulse force while jumping as while extending the legs during swimming. The techniques are also slightly different, as a jumping frog will not generate the most force until its legs are almost fully extended, whereas a swimming frog generates the most force earlier while its legs are still only partially extended.[21]

One type of bat can run as well as fly. While a few species of bat can laboriously crawl along the floor using their wings and tiny back feet, the **common vampire bat** (*Desmodus rotundus*) is capable of not only walking forwards, but sideways and backwards too. What is more, the bat can run at speeds approaching up to 2 metres per second, a trick that might literally help it run rings around the prey animals from which it sucks blood and feeds.[22]

Contrary to popular belief, the legs of other bat species are actually strong enough for them to walk. When attempting to crawl along the floor a clumsy **insectivorous bat** (*Pteronotus parnellii*) will exert larger forces with its hind limbs than the similar-sized

vampire bats (*Desmodus rotundus* and *Diaemus youngi*), showing that its supposedly spindly limbs are robust enough for crawling. Why most bats do not prefer this type of locomotion along the ground is unclear.[23]

Some monkeys leap from tree trunk to tree trunk, whereas others do not. For example, several New World primates including the **tamarins** and **marmosets** (*Cebuella pygmaea*, *Saguinus fuscicollis*, *Saguinus tripartitus* and *Callimico goeldii*) leap from trunk to trunk, whereas the closely related **marmosets** of the genera *Callithrix* and *Mico* do not travel this way, they climb from tree to tree rather than leap.[24]

Monkeys do not use a single technique for swinging through the trees; they use two distinct gaits analogous to walking and running. When moving slowly, a process known as continuous brachiation, a monkey will 'tree-walk' by holding on to a handhold support from the beginning of the swing until they grab hold of the next branch at the end of the swing. When moving fast, a process known as ricochetal brachiation, a primate will 'tree-run', with each handhold being interrupted by a period of ballistic flight from branch to branch. To do this, a monkey swings its arms at a greater speed, and each swing generates an additional vertical acceleration that helps fling them to the next branch.[25]

Orang-utans (*Pongo pygmaeus abelii*) and gibbons climb differently from African apes, the **gorilla** (*Gorilla gorilla gorilla*) and **bonobo** (*Pan paniscus*). Orang-utans and **gibbons** (*Hylobates concolor gabriellae*) take longer strides, moving their arms and legs a greater distance during each climbing step, and rotate their major joints over a greater range of motion.[26]

Chameleons (*Chamaeleo calyptratus*) take longer strides when climbing along a branch than other climbing lizards.[27]

Geckos (*Uroplatus*, *Palmatogecko*, *Stenodactylus*, *Tarentola*, and *Eublepharis* spp.) subjected to zero gravity will naturally adopt a sky-diving posture, believing they are in free fall.[28]

When birds bob their heads, they are actually keeping their heads still allowing them to fix their gaze on an object. The head of a bird will naturally move around as it walks, but by bobbing a bird can counteract this action as it takes each new step. For instance, a foraging **whooping crane** (*Grus americana*) bobs its head at a frequency relative to its walking speed that allows the bird to immobilise its head for 50 per cent of the time.[29]

The 35-tonne **sauropod dinosaur** (*Apatosaurus*) could run at an estimated 9 metres per second, compared to 6.8 metres per second for a modern elephant.[30]

The elusive **paradise tree snake** (*Chrysopelea* spp.) is the only vertebrate that is able to fly despite not having any limbs. The snake is a true glider, defined as covering a greater horizontal distance than it falls vertically, as opposed to a parachuter, which falls further than it travels. First the tree snake leaps from a tree into the air, and adopts a classic S-shape, which essentially changes the snake's body shape from that akin to a spear to one that is similar to a biplane, dramatically increasing lift. Then very quickly the snake undulates its body, and while no one can be sure why the snakes do this, the changing posture might serve to move the snake's centres of gravity and flow of air pressure in a way that allows controllable flight. The snake is so good at gliding that it can achieve a glide angle of just 13 per cent to the horizontal (90 per cent constituting free fall) and cover distances of 20 metres. From an aerodynamic point of view the paradise tree snake flies better than even a flying squirrel.[31]

Hummingbirds are the smallest flying vertebrates, and the only ones capable of sustained hovering flight. Hummingbirds can also fly backwards.[32]

Worm-like amphibians called **caecilians** are the only vertebrates known that use their whole bodies as a hydrostatic system for locomotion. Caecilians have unusually oriented body-wall muscles, an array of helical tendons that surrounds their body cavity and a vertebral column that moves independently of the skin. These odd adaptations allow them to use hydrostatic pressure within their whole bodies to burrow through soil. Using this system, for example, the Central American *Dermophis mexicanus* can generate approximately twice the maximum force in pushing itself forward as a similar-sized burrowing snake that must rely solely on muscles running longitudinally along the length of its body wall and vertebral column.[33]

Left or right

❧

I̲T̲ I̲S̲ D̲I̲F̲F̲I̲C̲U̲L̲T̲ T̲O̲ imagine how a snake can either be right- or left-handed. But, surprisingly, these limbless creatures do use one side of their bodies to greater effect than the other, during reproduction at least. Male **red-sided garter snakes** (*Thamnophis sirtalis parietalis*) have two penis-like organs used to inseminate females, both of which incorporate a complete reproductive system of hemipenis, testes, ducts and parts of the kidney. The right-sided reproductive organ is much bigger than that on the left, and produces a much larger ejaculate.[1]

Walruses (*Odobenus rosmarus*) tend to have longer right flippers than left, and prefer to use their right flippers more often. When feeding, walruses use their flippers to stir up the sediment of the sea floor in the search for bivalve shellfish. Not only do they use their right flippers more, but an examination of museum walrus skeletons has shown that, on average, the length of the scapula, humerus, and ulna bones in the right flipper are significantly longer than those in the left. Walruses are the first pinnipeds – the group of animals that includes seals, sea lions and walruses – known to prefer to use one side of their bodies more than the other.[2]

Racehorses are either left- or right-handed, or legged, depending on your point of view. Owners, jockeys and trainers have always suggested that their steeds often perform better turning in one direction over another, but now it has been shown that stallions generally prefer their left side, while mares prefer their right. Knowing a horse's stronger side could help bookmakers and punters alike as the direction of each race, and the bends involved, will suit some thoroughbreds better than others.[3]

Octopuses favour one eye over another. The creatures probably prefer to use one eye over another when peeping out of small hideouts and cavities looking for predators, or when stalking prey. It is the first time handedness – preferring one side of the body over another – has been seen in invertebrates. However, unlike in people, the majority of whom are right-handed, octopuses on the whole do not prefer one side over another.[4]

First-born **chimpanzees** (*Pan* spp.) tend to be left-handed, whereas chimps born later to the same mother tend to be right-handed. The situation again reverses itself after more than six offspring have been born, as the seventh baby onwards tends once more to be left-handed.[5]

Female **chimpanzees** (*Pan* spp.) and **gorillas** (*Gorilla* spp.) generally prefer to cradle their babies in their left arms than in their right.[6]

In the blood

Domestic cats have three different types of blood group, A, B and AB. But they lack the equivalent of the human blood group O. By far the most common blood type in domestic cats is blood type A, but the prevalence of different blood types varies with geographical location. For instance, all cats in Finland have blood type A, whereas only three-quarters of Australian cats are blood type A. The prevalence of different blood types also varies with breed.

COUNTRY	PERCENTAGE OF CATS THAT ARE TYPE A	PERCENTAGE OF CATS THAT ARE TYPE B
Finland	100	0
USA	99	1
England	97	3
Germany	93	7
Italy	89	11
France	85	15
Australia	73	26

table source [1]

PURE CAT BREEDS IN USA	PERCENTAGE FREQUENCY OF TYPE B CATS
Siamese	0
Oriental Shorthair	0
Burmese	0
Tonkinese	0
Russian Blue	0

PURE CAT BREEDS IN USA	PERCENTAGE FREQUENCY OF TYPE B CATS
Maine Coon	Less than 5
Norwegian forest cat	Less than 5
Abyssinian	5 to 25
Himalayan	5 to 25
Birman	5 to 25
Persian	5 to 25
Somali	5 to 25
Sphinx	5 to 25
Devon rex	25 to 50
British shorthair	25 to 50

table source [2]

Blood type in **cats** has no effect on coat colour or gender. But it can affect the survival of kittens. For instance, if a female cat of blood type B is bred with a male of blood type A, between 75 per cent to 100 per cent of their offspring will be blood group A. But if these kittens nurse milk from their mother within 24 hours of birth they may die. This is because kittens of type A or AB that nurse from a type B mother ingest antibodies against type A blood in their mother's milk. About a day after ingestion, these antibodies are absorbed by the kitten's stomach and gastrointestinal tract. Hemolysis, or the breaking open of blood cells, ensues, which can be fatal. A vet can tell when this may have happened to an otherwise healthy-looking kitten by looking at the cat's tail. Necrosis at the tip of the tail is a sign that hemolysis has taken place. However, after 24 hours have passed, antibodies do not cross the kitten's intestinal lining and the youngster can safely drink its mother's milk.[3]

Dogs have seven confirmed blood groups, although 20 separate blood groups have been reported in dogs at one time or another. Canine blood groups were once described using letters of the alphabet, including A, B, C, D, E, and beyond. Nowadays they are written as DEA1, DEA2, DEA3, and so on. Each canine blood group also comprises three subgroups, such as DEA1.1, DEA1.2, DEA1.3, and so on, and the number of canine blood groups makes blood transfusions between dogs extremely difficult. No systematic study of the prevalence of canine blood groups has been performed but over 60 per cent of native Japanese breeds such as the Shikoku and Akita have blood type DEA3. Non-native breeds such as the English setter, Maltese and beagle were also predominantly DEA3. However, DEA3 was not found in 122 Labrador retrievers tested in one study in Australia.[4]

The **European** or **domestic ferret** (*Mustela putorius furo*) is unique among mammals because it appears to have no blood groups. Every mammal so far studied has antigens for more than one type of blood, except domestic ferrets which have antigens for just one.[5]

There are an estimated 14,000 species of **insect** that feed on the blood of other animals and 300 to 400 of these regularly feed on the blood of people, or our livestock and pets.[6]

Smelly world

∾

A MALE **Asian elephant** (*Elephas maximus*) needs to wear the right perfume to attract the most females. Each year, males become sexually active, and enter a phase known as musth. As well as heightened sexual activity and aggression males produce a pungent cocktail of chemicals to advertise their mating status. Younger, less dominant bulls produce a mixture that contains predominantly one of two chemical forms of a pheromone called frontalin. Older bulls produce a blend containing equal amounts of both forms of the pheromone, a particular brand of scent that is more attractive to potential mates.[1]

Domestic **dogs** can determine the age, gender, sexual receptivity, and exact identity of another dog simply by sniffing the scent marks it leaves behind.[2]

Rats can smell a cat. The mere whiff of cat odour causes a rat's blood pressure to rise dramatically, even if it has never encountered a real cat before in its life.[3]

Birds recognise each other by smell just as mammals do, a finding corroborated only in 2004, when it was discovered that a small nocturnal Antarctic sea-bird, called a **prion** (*Pachiptila desolata*), uses smell to locate its mate in the dark.[4]

Nesting **birds** change the way they smell so that predators cannot sniff them out. Usually ground-nesting birds and ducks coat their feathers using a substance known as a preen wax, which repels water and inhibits the growth of feather-degrading bacteria.

However, when the birds begin incubating their eggs, they change the composition of this wax from compounds known as monoesters to others with a different structure called diesters. Birds coated with diester-based preening wax appear to be less easy to detect by predators relying on smell. Once the chicks have hatched, the birds return to preening with the old, more smelly wax.[5]

Male **sac-winged bats** (*Saccopteryx bilineata*) get their name because they have sacs on their bodies that they fill with urine, saliva and secretions from their genital region. The males use these so-called propatagial sacs to produce sexual odours that serve as a come-on to awaiting females. The males also use their wing tips to coat the fur on their backs with saliva, which they then use to scent-mark territory boundaries.[6]

Rats smell in stereo. With just one sniff, the rodents can work out the direction a smell is coming from. That is because the waft of an odour reaches each nostril 50 milliseconds apart, a tiny but significant difference that allows the **rat** (*Rattus norvegicus*) to work out from where the smell is emanating. When one nostril is blocked, a rat's ability to sniff out the direction of a smell is greatly reduced.[7]

The skin of some Australian frogs contains a natural mosquito repellent. The **frog** species *Litoria caerulea*, *Litoria rubella* and *Uperoleia mjobergi* all secrete compounds that deter mosquitoes from biting. The nature of the repellent is unknown but the frogs are likely to secrete volatile compounds including terpenes, which they might obtain from their food.[8]

Wasps and **bees** have such good noses that they are being used to quality-check the freshness of supermarket food. They are also being trained to sniff out the chemical ingredients of bombs or the odours produced by people with certain diseases.

For instance, **honeybees** (*Apis mellifera*) have been used to gauge the quality of supermarket strawberries. When the bees expect to encounter suitable food, they automatically stick out their proboscis to sample the scent it produces. This talent has been exploited by training honeybees to stick out their proboscises when they are presented with strawberries that are ripe enough for sale in supermarkets. They have even been trained to sniff out illicit stashes of tobacco and cheap, fake whisky. Honeybees and a species of wasp called *Microplitis croceipes* have also been trained to sniff out illicit drugs and chemical explosives, being as talented at the work as sniffer dogs.[9]

US army researchers have used **bluegill sunfish** (*Lepomis macrochirus*) to warn them when drinking-water supplies have become contaminated. Electrodes are implanted in the fish, which are linked to a computer. When the fish are set free, the computer then records their behaviour. If the fish encounter toxins in the water, their breathing changes – they breathe faster and deeper, and cough more. They also begin to swim in an erratic way. If enough of the fish begin to display these odd behaviours the computer sounds an alarm.[10]

Zoological definitions

Allogrooming	The act of an animal grooming another of the same species
Anetrus	When a mammal is not sexually active during its reproductive cycle
Axilla	The angle between a forelimb and the body, the equivalent of the human armpit
Baculum	A bone found in the penis of some mammals
Brachiate	To use the arms to swing between branches
Bruce Effect	Happens to mice. When a new male arrives, a female undergoing the Bruce Effect will abort her foetus and again become sexually receptive
Consort	A female monkey that forms a temporary relationship with a male to ensure she has an available partner when she is ready to mate
Coolidge Effect	The ability of male hamsters efficiently to choose new female sexual partners over old ones
Deme	A small group of animals that randomly breed with one another
Durophagy	The eating of shells
Flehmen	The expression on the face of an animal sniffing a strong odour. Often characterised by a wrinkled nose, retracted lips and clenched teeth

Flense	To strip the blubber from a seal or whale
Iteroparous	Regularly has offspring
Lanugo	The birthcoat of baby mammals later shed in adult life
Lobtailing	A whale beating the water with its tail
Lordosis	When a female arches her back to signal she is ready to mate
Myrmecophagy	Eating ants and termites
Nares	External nostrils
Os clitoris	A bone found in the clitoris of some mammals
Pelage	All the hairs on an individual animal
Pronking	An animal vertically jumping on the spot, so all four feet leave the ground
Rhinarium	The moist skin around the nostrils of many mammals
Stotting	Another word for pronking
Tine	The point on an antler
Volant	Being capable of powered flight

table source [1]

The power of the goat

∿

THE GOAT WAS SACRED to the Sumarian and Babylonian gods, Ningirsu and Marduk, whereas the Pharaonic god Osiris appears sometimes in the form of a goat. As an infant, Zeus – the father of gods and men – was fed milk of the goat Amalthea, the horn of which has been considered a symbol of fertility and abundance.

In oriental cultures, the Chinese goat spirit Yang Ching exists. The goat is also a Mongolian god, whereas Russia has a wood spirit, the *Leshi*, which resembles Pan and the Satyrs in the shape of a human with the horns, ears and legs of a goat. In India, the term 'goat' also means 'not-born'; and is thus the symbol of the unknown primary substance of the living world.

The milk of goats and the quality of the meat are important to the Hebrews, as mentioned in the Book of Genesis. In Egypt, Pharaoh Tutankhamen ordered 22 tubes of his favourite goat-milk cheese to be placed in his tomb so that he could not only be nourished as he travelled forward to the next world but also so that he could offer them as gifts to the gods.[1]

Ripe for mummification

CATS, BIRDS, MONKEYS, gazelles and even crocodiles were mummified by the ancient Egyptians. It is likely that cats were selectively chosen to be mummified, as mummified cats have been discovered with skeletal damage that suggests their necks were deliberately snapped. They were also mummified with the same level of care and attention accorded to people. The Egyptians used sugar gum, animal fat, plant oil, pistachio resin, bitumen, beeswax, balsam and cedar resin to embalm their pets, substances similar to those found in human mummies from the same period, between 818 BC and 343 BC. Falcons, which were thought to have been kept as pets, were also lavishly mummified with great care, while ibises and hawks, which may have been offerings to the moon god Thoth, were mummified in a more casual manner.[1]

Laugh and play

∾

LAUGHTER AND JOY are not exclusively human traits.

Our closest relatives, **chimpanzees**, **gorillas**, and **orang-utans**, utter laugh-like sounds when they are tickled, a phenomenon first reported by Charles Darwin in 1872.

Chimpanzees, for example, regularly play-pant when playing with each other, and have at least three distinct playing routines:

- **Mouthing** The performer puts its open mouth against the target's body and nibbles. This is thought to be a kind of tickling among chimpanzees.

- **Tickle** The performer presses the fingers of one or both hands on the target's body. This is often accompanied by a succession of alternating flexing and stretching finger movements.

- **Chase** The performer runs after the target or follows the target at walking speed, during which the performer often reaches for and tries to catch the target by the leg. Even when chasing around a tree trunk or shrubs, one of the players often approaches the other from behind and tries to catch it.[1]

Rats will vigorously play with one another, whereas guinea pigs do not.[2]

When young **rats** play they emit a specific sound, chirping at a frequency of 50 hertz. They emit the sound even more strongly when tickled on their stomachs. Just like humans who are more ticklish on certain areas of the body, young rats are most ticklish around their 'tickle skin', an area concentrated at the nape of the neck.[3]

When it comes to playing, the size of your brain counts for a lot. Generally speaking, play behaviour is more common within those classes of vertebrates that have bigger brains. Those groups of **birds** and mammals with the largest brains also contain more playful species. When it comes to sexual play, a region of the brain called the amygdala is vital. Animals with a large amygdala are more likely to engage in play with each other in a sexual way, possibly because the amygdala is critically involved in the learning and regulation of fear. If two strangers of opposite sexes meet, they will naturally be hostile to each other, despite the potential for sexual attraction. Play helps overcome this tension.[4]

Amazing fakes

∾

ONE OF THE GREATEST zoological fakes was uncovered in 2001 after having duped biologists for 80 years. At stake was the rare discovery of a large new mammal, a supposed new species of ungulate known by locals in south-east Asia as either the linh duong, or mountain goat, or the khting vor, the wild cow with liana-like horns.

In 1929, a number of sets of elaborate horns of this remarkable animal, ringed with ridges and curling backwards towards the tip, were discovered, but no one knew quite what to make of them. Experts of the day decreed they belonged to the **kouprey** (*Bos sauveli*), an extremely rare bovid living in Indochina, and they were deposited at the Kansas Museum of Natural History in the USA.

Then in 1993, similar horns began turning up regularly in the food markets of Vietnam and Cambodia. They looked unlike any seen before and convinced scientists that this was a new species of large mammal, a jungle-living ungulate which they named *Pseudonovibos spiralis*. Oral evidence from local hunters suggested that it was a buffalo-like but very agile animal, and conservationists frantically added the new mammal to world lists of endangered species, whereas experts debated whether it was more closely related to sheep, goats or cattle.

But in 2001, a detailed forensic analysis of the horns' structure and DNA revealed the truth, highlighting the pitfalls that go with describing a new species without having a whole specimen. The horns, it turned out, were those of a **cow** (*Bos taurus*), which had been skilfully heated, squeezed, bent and carved by local craftsmen. Rather than being a deliberate fraud, the practice of adulterating cow horns in this way may go back centuries, as pictures of similar horns have since been found in illustrations within a Chinese encyclopaedia dating from AD 1607.[1]

Some of the most exciting fossil discoveries have turned out to be elaborate fakes.

One such forgery was a fossil fly that had been regarded as one of the greatest treasures of the Natural History Museum in London. The fly was almost perfectly preserved in amber, and was thought to represent the earliest known member of the family of insects that include the ubiquitous housefly. The supposedly 38-million-year-old fossil, of a **latrine fly** called *Fannia scalaris*, also showed just how little our house pests had evolved over time, as similar creatures were known to live in cities throughout Europe during the 19th century.

The amber fossil was first documented in 1850, and was acquired by the museum 72 years later. But it took another 71 years before its true provenance was known. It was only when an ancient-insect expert studied the fossil in 1993 that he discovered the amber it was encased in had been cut in half, and a modern fly inserted, before the two pieces were glued back together. The identity of the Victorian forger remains a mystery.[2]

A fossil that was heralded as a missing link between **dinosaurs** and **birds** was also finally unmasked as being a fake, leaving palaeontologists around the world with egg on their faces. The feathered creature was given the scientific name of *Arachaeraptor liaoninggensis* and had a bird-like body, while sporting a straight, rigid tail that looked like that of a fleet-footed predatory dinosaur known as a dromeosaur.

This blend of dinosaur and bird features was such an exciting, rare discovery that it was announced at a press conference by the National Geographic Society in Washington DC in October 1999.

Later it emerged that the fossil was the work of forgers looking to sell it for profit. A study using X-rays revealed the fossil was a patchwork of different animal parts glued together. To a fossilised bird's body the fakers had glued up to four different bits of dinosaur tail. Embarrassed experts, however, still think the front part of the fossil may yield some interesting insights into bird evolution. The fossil impression shows an animal that looks as if it could have flown better than *Archaeopteryx*, which is famously the world's oldest bird yet known.[3]

What was thought to be the world's largest ever **spider** is nothing of the kind. A fossilised specimen of the supposed spider, a species called *Megarachne servine*, was dug up in 1980 from the Bajo de Véliz Formation of San Luis Province, Argentina. It looked remarkably like a giant tarantula, and with a body length of 339 millimetres would have been the largest known spider ever to have lived on earth. However, a row over ownership led to the specimen being locked in a bank vault, where it could not be studied. It was not until 2005 that scientists confirmed, having studied a recently discovered second specimen, that M. *servine* is not a spider at all, but instead is a bizarre example of another type of arthropod known as a sea scorpion.[4]

Mistaken identity

❧

FLYING LEMURS ARE not related to lemurs. And they do not fly. The animals, also known as colugos, have lemur-like faces and bodies but belong to the completely separate order *Demoptera*. They are also gliders, throwing themelves into the air and gliding from tree to tree using a large, furry membrane extending from the sides of the neck to the forepaws, and from the forepaws back to the hind feet and end of the tail. Similarly, **flying squirrels** (*Glaucomys* spp.) glide rather than fly. Bats are the only mammals to have achieved true flight.[1]

The **red panda** (*Ailurus fulgens*) is not a panda. Discovered in 1827, before the **giant panda** (*Ailuropoda melanoleuca*) in 1869, the red panda has long been considered a close relative of its larger namesake. Both animals even share a unique pseudo 'thumb'. But the red panda is actually either a type of raccoon – being placed within its own subfamily, *Ailurinae*, within the raccoon family, the *Procyonidae* – or it is a member of the wider mustelid group *Musteloidea*, which includes raccoons, skunks, weasels and otters.[2]

Killer whales (*Orcinus orca*) and **pilot whales** (*Globicephala* spp.) are actually dolphins. Growing to a length of almost 10 metres, killer whales are the largest member of the **dolphin** family *Delphinidae*.[3]

The **two-toed sloth** (*Choloepus hoffmanni*) actually has three toes, fused together. Its name drives from its two claws.[4]

Hyenas (*Hyaenidae*) may look a lot like dogs, but they are more closely related to the **cats** (*Felidae*), **civets** and **genets** (*Viverridae*) and **mongooses** (*Heprestidae*).[5]

The **kinkajou** (*Potos flavus*) is a furry omnivore that feeds off sweet fruits, honey, flowers and small insects and other invertebrates. It also has a prehensile tail and inhabits the trees of South and Central America. No wonder then that it is referred to by local people as *mono de la noche*, or monkey of the night. The description is apt, for when it was first discovered by taxonomists they classified it as a primate. However, the description could not be more wrong. Genetic evidence has conclusively proved that the kinkajou is in fact an omnivore closely related to the raccoon.[6]

Mega bats, the group that includes the larger fruit-eating bats such as flying foxes, were once classified as flying primates.[7]

The rarest species of dog in the world may not be a unique species at all. The **red wolf** (*Canis rufus*) of the coastal plains and forests of the south-eastern USA may well be extinct in the wild, and huge efforts have been made to breed the few survivors that exist in captivity. However, there is evidence that the red wolf may actually be a hybrid between **grey wolves** (*C. lupus*) and **coyotes** (*C. latrans*). Whereas morphological studies of the animals' skulls and genetic evidence of mitochondrial DNA point to a unique origin, other genetic evidence suggests that they were simply born from the relatively recent breeding of the grey wolf and coyote. Some experts even say that the surviving red wolves may be a mix of pure greys, pure reds, red wolf/coyote hybrids and red wolf/grey wolf hybrids.[8]

The story is further complicated by recent suggestions that the grey wolf and the domestic dog are in fact one and the same species and should both be described as C. *familiaris*.

There is also debate over whether any pure-bred **dingoes** remain in Australia, or whether all the animals currently described a dingoes have actually interbred with domestic dogs to such an extent that it is impossible to tell the difference. The same argument is also made regarding the British wild cat, with many experts arguing that they are no longer a distinct species but are simply the result of generations of breeding with domestic feral tabbies.[9]

Musk deer are not deer. Rather than belonging to the deer family *Cervidae*, **musk deer** are the only living members of a separate family of mammals called *Moschidae*.[10]

Bees, wasps and even snakes, to name but a few, all do it. But butterflies do it best. These animals have an extraordinary ability to mimic the colourful patterns of one another, so that one species will evolve an almost identical wing pattern to a second species despite the fact that the two species are unrelated. The best example can be found within the *Heliconius* butterflies of South America. One species, *H. erato*, has evolved into almost 30 different races across its range, with each having a different wing pattern. But the unrelated species *H. melpomene* is not to be outdone. It has also evolved almost 30 races, again each with a different wing pattern. And every *H. melpomene* race is an almost identical mimic of its corresponding *H. erato* neighbour. Both butterflies are distasteful to birds and their wing patterns serve to warn any predators that they are unpalatable. Biologists think the two species mimic each other to increase the impact of this message, as two species sharing the same wing pattern doubles the chances that a predator will have come across the warning signal before and learnt through experience that this is not a butterfly worth eating.[11]

Moths and butterflies often come in two different forms depending on the time of year. In the summer a species of butterfly may appear with wings of a certain colour and pattern, whereas in the winter later generations of the same species may look completely different. The phenomenon is known as seasonal polyphenism. For example, tropical butterflies typically look different during the wet and dry seasons, and **satyrine butterflies** such as *Bicyclus anynana* appear more cryptic during the dry season and have larger eyespots on the wing in the wet season. The most extreme form of seasonal polyphenism among temperate butterflies is shown by the **map butterfly** (*Araschnia levana*). Spring-generation butterflies are red and white, whereas the summer-generation butterflies are black and white.[12]

Aquatic insects, such as **beetles**, dragonflies and true bugs have a strange attraction towards red and black-coloured vehicles.

Often, such insects will swarm above and land on the boots, bonnets and roofs of red and black cars, while female insects often lay their eggs en masse on the surfaces of such vehicles.

It is likely that the insects are being fooled by the reflective and polarisation characteristics of the car bodies. Aquatic insects detect water on the basis of the horizontal polarisation of light reflected from water surfaces. Red and black roofs and bonnets probably produce a similar visual signal that fools the polarisation-sensitive visual system of these insects. In large parking lots, the visual deception can increase significantly, because of the added polarising effects of many cars parked close to each other. It has even been suggested that people should choose other more environmentally friendly colours for their cars to avoid insects becoming confused and laying precious eggs on vehicles where they will not survive.[13]

Poison-dart frogs (*Phyllobates* spp.) raised in captivity are not poisonous. This is because the frogs obtain their poison in the wild by eating small beetles of the genus *Choresine*, which contain high levels of neurotoxic chemicals called batrachotoxins. These chemicals are secreted through the frog's skin.[14]

Male and female **chimpanzees** have markedly different faces, but male and female gorillas do not. As in people, male chimps with broad faces and enlarged cheekbones are perceived to be more attractive by the opposite sex, and because of this selection pressure male chimps have evolved wider faces than females. Gorillas of both sexes have similar-shaped faces.[15]

Up to three-quarters of the fish sold as red snapper in US fish markets over the past few years are not red snapper at all. According to the US Food and Drug Administration, the only fish that can be legally sold in the USA as **red snapper** is the species *Lutjanus campechanus*. But genetic tests have revealed that between 60 per cent and 94 per cent of fish sold as red snapper in the US are actually **lane snapper** (*L. synagris*) or **vermilion snapper** (*Rhomboplites aurorubens*).[16]

A **rat** destined for the kebab skewer caused consternation among biologists when in 1996 it was discovered for sale, in a market in Laos, on a table next to some vegetables. The rock rat, or *kha-nyou* as it is known to locals, was given the species name *Laonastes aenigmamaus*, which translates as 'stone-dwelling enigmatic mouse'. The animal looked like a cross between a large dark rat and a squirrel, and was so bizarre that it was given its own family, the *Laonastidae*, the first new mammal family created since 1974. However, 10 years later, scientists realised that the enigmatic animal did not represent a unique new family of mammals after all. Instead it was the last known surviving member of a group of mammals that lived 11 million years ago.[17]

Telling look-alikes apart

ॐ

Seals and sea lions

True **seals** (*Phocidae*) lack external ears, and have a thick layer of blubber. Phocids also have hind flippers that point backwards and cannot be turned forward to help with walking on land.

Sea lions and **fur seals** (*Otariidae*), in contrast, have external ears, and have less blubber, although they compensate for this by generally having thicker fur. Otariids can bring their hind flippers forward, allowing them to walk relatively well out of the water.

The **walrus** shares characteristics of both, as it lacks an external ear but is capable of bringing its hind flippers forward. It is also the only pinniped to have descended testes, and belongs within a family of its own, the *Odobenidae*.[1]

Alligators and crocodiles

The family *Alligatoridae* include the **alligator** and **caiman**. They have wide, rounded U-shaped snouts built for strength, capable of withstanding the stresses caused when the jaws crack open turtles and hard-shelled invertebrates, which form part of their diet. The teeth in the lower jaw are almost completely hidden when the mouth closes, fitting neatly into small depressions or sockets in the upper jaw. Alligators and caiman lack salt glands on their tongue and have sensory pits around their jaws, which are capable of detecting differences in water pressure.

True **crocodiles** belong within the *Crocodylidae*. They tend to have longer and more pointed V-shaped noses that are relatively weaker, designed to catch a wider variety of prey. The large fourth tooth in the lower jaw sits outside the upper jaw when a crocodile's mouth is closed and is easily visible. Crocodiles have salt glands on their tongue and pressure-sensing pits all over their bodies.

One crocodilian, the **gharial** (*Gavialis gangeticus*), belongs in a family if its own, the *Gavialidae*. It has a long, thin snout uniquely adapted to catching fish.[2]

Dolphins and porpoises

They may look the same, but in terms of anatomy and behaviour, **dolphins** and **porpoises** are as different as cats and dogs. Porpoises never grow larger than 2.5 metres in length and lack the distinctive beak or rostrum of dolphins. They tend to wander the oceans alone rather than in large social groups like dolphins and they do not cooperatively hunt together. Unlike their larger cousins, porpoises do not form long-lasting social bonds.[3]

Butterflies and moths

There is no specific list of attributes that can distinguish all butterflies from all moths. But as a general rule:

- Butterflies have knobbed antennae; those of moths range from straight filaments to feathery or branched.

- Butterflies have smooth, slender bodies; moths tend to be plump and fuzzy.[4]

- Most butterflies fly during the day; most moths fly at night.

- Butterflies generally rest with their wings held upright; moths spread them out.

- Most of the brightly coloured Lepidoptera are butterflies.

Alliterative collective nouns

army of ants
caravan of camels
clowder or cluster of cats
dule of doves
flamboyance of flamingos
gaggle of geese
husk of hares
herd of horses
leap of leopards
lounge of lizards
pandemonium of parrots
pride of peacocks
pod of pelicans
prickle of porcupines
rhumba of rattlesnakes
school of sharks
scurry of squirrels
sounder of swine
wisdom of wombats
zeal of zebras [1]

What's in a name?

∽

THE ENGLISH-SPEAKING world did not have a word for **shark** until just 500 years ago. The word shark derives from the Mayan word *xoc*, and first appeared in the 16th century. For one hundred years prior to that, English speakers used the Spanish word *tiburón*, which in turn was borrowed from the Carib Indians. The reason is thought to be because, whereas large sharks were known to the Greeks and Romans, and references to large sharks are found in the writings of many classical writers, medieval people in Europe rarely, if ever, encountered the beasts. They only fished close to shore, whereas large European sharks do not venture inland up rivers and streams. It was only when explorers reached the American tropics that sharks, and the fear of them, entered the European psyche.[1]

The zoological name for the **walrus** (*Odobenus rosmarus*), literally means 'toothwalk', deriving from the Latin words *odontos* for teeth and *baenos* for walk. Walruses are so called because of their habit of using their long downward pointing tusks as a fifth limb, striking the points into the ice to help haul their huge frames out of the water.[2]

More than one name to call a cat

THE **mountain lion** (*Pumas concolor*), a single species of large cat living across the Americas, has more colloquial names than any other mammal in the world, including cougar, Colorado cougar, Yuma cougar, eastern cougar, catamount, panther, Florida pan-ther, American lion, Mexican lion, silver lion, gray lion, moun-tain lion, swamp lion, plain lion, leon americano mountain screamer, mountain tiger, puma, Wisconsin puma, painter, deer tiger, deer cat, king cat, sneak cat, Indian devil, purple feather, cuguacuara, cougouar, cuguacuarana, carcajou, Silberlöwe léon, léon colorado, léon de montaña, léon sabanero, leon bayo, onça vermelha, onça parda, suçuarana, quinquajou, tigre rouge, tygre, guasura, yaguá-pytá, cabcoh, leopardo, reditigri, catawampas, long tail, pampas cat and swamp devil.[1]

That time of life

❧

A GRANDMOTHER **lowland gorilla** (*Gorilla gorilla gorilla*) has been seen teaching her daughter how to behave as a mother to her newborn baby. The incident took place at San Diego Wild Animal Park on 30 October 2000, when an 11-year-old female, Ione, gave birth to her second baby, a male. Ione abandoned her child on the floor. Ione's mother, a 21-year-old gorilla called Alberta, was then seen repeatedly picking up the child, moving it closer to Ione, and occasionally placing it in Ione's arms. Alberta continued to teach her daughter mothering skills throughout the first 4 days after birth, over which time Ione's maternal behaviour gradually improved.[1]

Female **orang-utans** (*Pongo abelii*) do not undergo the menopause, despite living to well over 50 years of age.[2]

Male monkeys have sympathetic pregnancies just like their human counterparts. Male **common marmosets** (*Callithrix jacchus*) and **cotton-top tamarins** (*Saguinus oedipus*) put on an extra 10 per cent of their body weight over the course of their mate's pregnancy. However, males put on the extra weight midway through the gestation period whereas females put on weight later as the baby nears full term.[3]

Male **chimpanzees** (*Pan troglodytes schweinfurthii*) grow quickly until they are 13 years old, reach their peak weight around 24 and then become progressively lighter as they age further. Females grow quickly until about the age of 11 and their weight stabilises at 21. When given an adequate provision of bananas, for example, wild chimps' body weights increase by an average of 17 per cent.[4]

Age withers us all, and **baboons** are no different. On average, one in seven of all the cells in a baboon's skin will have lost the ability to divide by the time the animal reaches the age of 30 years.[5]

A toddler **chimpanzee** that is just three years old is able to recognise when an adult, even a complete stranger, needs help performing a task and will assist them. They will do so even when there is no benefit to themself, such as gaining a reward or being praised. Such altruistic behaviour, called instrumental helping, has been recorded in only one other species of infant animal: human toddlers.[6]

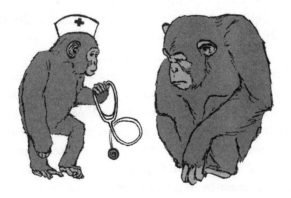

Baby marsupials, which are born immature after a short gestation period compared to other mammals, have an undeveloped immune system and are dependent on their mothers for immunological protection. Young **tammar wallabies** (*Macropus eugenii*) get around this problem by deriving much of their immunity to disease and bacteria in their mother's pouch from antibodies secreted in their mother's milk, which is also thought to contain antimicrobial proteins. A female tammar wallaby produces two different types of milk for its joey depending on its age: 'early' milk that has ingredients more suitable for young infants for the first 180 days; and 'late' milk, which is higher in some proteins, for older offspring for the remaining 150 days. A mother wallaby is also capable of producing both types of milk at the same time from adjacent teats: one to feed a young newborn, and the other to feed an older animal that has left the pouch.[7]

Once they are born, baby mammals are thought to take as much rest and sleep as possible, and gradually sleep less and less as they get older. Newly born **killer whales** (*Orcinus orca*) and **bottlenose dolphins** (*Tursiops truncatus*) are the exceptions. For the first month of a newborn's life mother and calf of both species take no rest at all. They do not sleep and remain mobile for 24 hours a day. As the calf grows older, both mother and offspring will gradually sleep more and more until they rest as often as other adults that do not have calves, which can be between five and eight hours a day. It is not clear how both species cope with these sleepless calf-rearing nights, as a lack of sleep would kill people, rats and even flies. But it is thought that staying awake, however it is done, helps the newborn calves to breach the surface regularly and breathe, avoid being caught napping by predators and maintain their body temperature until they have grown larger and put on enough insulating blubber.[8]

Newborn **laboratory rats** fall in and out of sleep incredibly rapidly. In fact, a two-day-old rat will sleep for just 15 seconds before waking for another five seconds. The rat then falls asleep once more for another 15 seconds, repeating the cycle.[9]

Birds and some species of mammal are the only modern animals to sleep resting on folded limbs. However, there is intriguing evidence that some dinosaurs also slept just as birds do. In 2004, a fossil of a tiny **dinosaur** called *Mei long* was discovered curled up in the stereotypical tuck-in sleeping posture of many living birds, with its head resting around the left side of its body alongside its elbow. The back legs of M. *long*, which was discovered in the Early Cretaceous Yixian Formation of western Liaoning, China, and thought to have lived between 128 and 139 million years ago, were also folded and tucked underneath its body, just as birds sleep or rest.[10]

Young tadpoles of the **frog** species *Xenopus laevis* spend 99 per cent of their time doing nothing.[11]

Tadpoles that mature the fastest do not necessarily grow into the most able frogs. Tadpoles of the **frog** *Rana latastei* that metamorphose into frogs earlier have shorter, stubbier leg bones and are not able to jump as far as frogs that metamorphose from tadpoles that took their time growing up.[12]

A worm-like amphibian, the **caecilian** (*Boulengerula taitanus*), takes parenting to a whole new level. Females look after their brood of young, but do more than just provide a meal of yolk in an egg. Instead, they offer their bodies.

By elongating specialised stratified epithelial cells, brooding females transform their skin until it is twice as thick as that of non-breeding females. The developing young then tuck into their mother's skin, using highly specialised teeth to peel off and eat the outer layers. Like the milk of mammals, the skin contains high levels of lipids, providing a rich supply of nutrients for the developing offspring. The practice, called dermatotrophy, is a highly unusual mode of parental care previously unknown in tetrapods.[13]

One species of **squid** does something no other squid does. *Gonatus onyx*, one of the most abundant cephalopods in the Pacific and Atlantic Oceans, carries its mass of brooding eggs within its arms as it swims. All other squid lay their eggs on the sea floor, abandoning them to their fate. The parenting skills of G. *onyx* come at a cost, however. Each little bundle of joy can contain between 2,000 and 3,000 eggs, and the size and weight of this extra burden makes swimming much more difficult, making it more likely that the relatively immobile caring parent will be eaten by a predatory whale or elephant seal.[14]

If you have ever wondered how many flowers a bee has to visit to obtain enough pollen to live and breed, we now have the answer. Of 41 bee species living in Europe, 85 per cent of them must visit 75 individual flowers or more to find enough food to raise just a single offspring. But the number of flowers that must be visited by individual bee species varies greatly. Some only have to visit 18 flowers to obtain enough pollen, whereas others might have to visit many more. One large and conspicuous European species of **bee** (*Megachile parietina*) must visit an average of 284.75 flowers to raise a single young. Considering that each female builds ten brood cells during her entire life, she must visit 28,475 flowers to obtain enough pollen to raise her offspring.[15]

Newly hatched hawkmoths seek out flowers that are exactly 32 millimetres in diameter. But what appears to be an overly fussy taste for nectar may actually save the young moth's life. Flowers that are much smaller than 32 millimetres do not offer much in the way of food, whereas more experienced adults home in on flowers of any size or colour. But large petals will fill the field of vision of a young **hummingbird hawkmoth** (*Macroglossum stellatarum*) confusing the insects as to how close they are. If they fly too near, they risk becoming trapped in the sticky interior of the flower.[16]

The **bee** species *Megachile parietina* builds its hive on rocks, making the individual cells of its hive out of mud.[17]

The **pygmy goby** (*Eviota sigillata*) lives for no more than 59 days, the shortest recorded lifespan of any vertebrate.[18]

There is one species of frog that produces tadpoles that never grow up. The **African clawed frog** (*Xenopus laevis*) sometimes produces tadpoles that grow into giants, without ever metamorphosising into adults. These grossly deformed giant tadpoles, which sport a hunchback and are on average four times longer and up to 50 times more massive than normal tadpoles, can survive for years, although they are not capable of reproducing.[19]

Naked mole rats (*Heterocephalus glaber*) live longer than any other rodent, with a lifespan of up to 28 years, seven times longer than mice of a similar size.[20]

Over the course of its life, the **blue whale** (*Baleanoptera musculus*), the largest animal known to have lived, will grow to weigh between 130 and 150 tonnes. That is roughly equivalent to the weight of 33 **African elephants** (*Loxodonta africana*) or 65 million **pygmy white-footed shrews** (*Suncus etruscus*), the smallest known mammal.[21]

Maggot larvae of the **fruit fly** (*Drosophila*) grow so quickly that they gain 7 per cent of their body weight every hour.[22]

Birds live longer than mammals of similar body size.[23]

Mute swans (*Cygnus olor*) that reproduce early in life also stop breeding early, and vice versa. Which pattern a swan adopts appears to be genetically inherited.[24]

Chicks of the **mallee fowl** (*Leipoa ocellata*) begin life by being completely buried alive. The chicks survive despite hatching within, and being completely encased by, the loose sand of their incubation mound. The chicks then have to burrow out of the mound to escape.[25]

Sick dads sometimes make better parents, especially when they fear they have not got long to live. When **male blue-footed boobies** (*Sula nebouxii*) reach the relatively old age of ten years and begin feeling ill, they suddenly turn into superdads investing so much energy into rearing their young that they increase their success rate at raising fledglings by almost 100 per cent. When younger males between the ages of three and nine years old fall ill, the number of their fledglings that survive falls by almost 20 per cent.

The older birds turn into superdads because it makes more sense for them to disregard their illness and invest more time and effort in caring for offspring during the little time they have left, increasing the chances of passing their genes on to the next generation. Young males put their energy into looking after themselves, figuring that it is better to survive over the coming years, and giving themselves a chance to raise a greater number of chicks over their lifetime.[26]

Old female **blue-footed boobies** (*Sula nebouxii*) feed their young at higher rates and lose more weight doing so than do middle-aged females, a further example that some animals increase their investment in reproduction as they near the end of their lives.[27]

Some birds intuitively know when their eggs are going to hatch. **Southern giant petrels** (*Macronectes giganteus*) periodically go away on foraging trips while incubating their eggs. But days before their chicks hatch, the parents dramatically reduce the time they spend away from the nest, seemingly possessing an internal clock that warns them that their young are about to emerge.[28]

There is one species of bird that uses temperature to regulate the sex ratio of its young, something previously only thought to be done by reptiles. The **Australian brush-turkey** (*Alectura lathami*) is known as a mound-building megapode because males throw huge piles of earth over each clutch of eggs. And the temperature inside dictates how many males or females will hatch. More males hatch at lower incubation temperatures and more females hatch at higher temperatures, whereas the proportion is equal at the average temperature found in mounds.[29]

Female **lesser black-backed gulls** (*Larus fuscus*) produce heavier eggs the more they interact with other birds. And the amount of yolk in the eggs goes up. No one knows why.[30]

Individual pairs of migrating male and female birds can arrive at their summer mating grounds at exactly the same time, despite overwintering in different places and setting off on their migrations at different times. Pairs of **black-tailed godwits** (*Limosa limosa islandica*) spend the winters apart, with the males and females often living in different locations separated by distances of up to 1,000 kilometres. Despite not knowing when their mate will set off, and never meeting them during their migration, male and female godwits will arrive at their summer breeding grounds within an average of just three days of each other, displaying a remarkable sense of timing considering that godwits continue to arrive at their breeding grounds over a one-month period in April. It is not clear how pairs of birds arrive synchronously, but it could be that each pair winters in areas of similar quality, which in turn means they will be in a similar condition to arrive at specific times in spring. Or it could be that they are genetically or physiologically primed to migrate at the same time, or they may independently synchronise their arrival to the optimal time for each specific breeding location.[31]

Adult **molluscs** give their kids a head start in life. Those such as gastropod snails often lace their spawn, which contains hundreds or thousands of embryos, with antimicrobial fatty acids. These protect the young against infection by common disease-causing marine bacteria.[32]

A pregnant **cow's** body temperature falls during the two or three days before it gives birth, and can be used to predict when the calf will be born.[33]

An average domestic **pig** (a Danish Landrace) will have a brain cortex at birth that contains 425 million neurons. However, another breed of pig, the diminutive Göttingen minipig, has just 253 million neurons at birth. That does not mean that the Danish Landrace is a cleverer animal. When the pigs grow into adults, the number of neurons in the brain cortex of the Danish Landrace hardly changes. However, the brain of the Göttingen minipig continues to develop for weeks or months after birth, and by the time it is an adult the number of neurons in its cortex has significantly increased to 324 million.[34]

Beavers living in northern latitudes are nocturnal during the summer months. But during the winter, when they live in burrows under the ice and snow where little light penetrates, the animals change their behaviour completely. The **beavers** (*Castor canadensis*) become free running, and follow a 27-hour pattern of behaviour. What is more, the whole family of up to nine individuals appear to make the switch together, synchronising their activities.[35]

Reindeer have an unusual body clock that does not rely on a usual 24-hour day and night cycle. All animals have body clocks that regulate physiology and behaviour during certain times of the day. But many **reindeer** (*Rangifer tarandus platyrhynchus*) live at extremely high latitudes, for instance in the Arctic archipelago of Svalbard. In summer and winter at these latitudes, the sun neither rises nor sets, with daylight being continuous in summer, whereas winter days are perpetually dark. During these extreme light conditions, the reindeers' body clocks subside, and they lose circadian rhythmic activity completely. The only time the reindeer behave according to a usual light and dark cycle is during the spring and autumn.[36]

The mammary glands of a female **pig** will warm up shortly before she gives birth, allowing the newborn piglets to find the gland and begin suckling soon after being born. After a short while, each piglet in the litter will choose a specific mammary gland to suckle from, which it will continue to return to. Those glands not chosen by a piglet will dry up. Domestic piglets prefer to suckle the front teats, while **wild boar** (*Sus scrofa*) piglets prefer hind teats. Piglets are born with eight teeth which are used to fight other piglets for access to their mother's milk, one of the few examples known of a mammal being born with weaponry that can be used to fight its newborn brothers and sisters.[37]

Male **red deer** (*Cervus elaphus*), **fallow deer** (*Dama dama*), **white-tailed deer** (*Odocoileus virginianus*) and **moose** (*Alces alces*) grow their antlers during spring or summer. **Roe deer** (*Capreolus capreolus*) grow their antlers in autumn.[38]

The brains of different **mammals** develop at different times of life. The brains of guinea pigs, sheep and monkeys undergo a growth spurt while the foetus is still in the womb. The brains of rats and rabbits, however, significantly develop only once they are born. Pigs are peculiar, in that, like humans, their brains develop perinatally, with a brain growth spurt extending from mid-gestation to early post-natal life.[39]

Female **mice** raised together with their biological brothers and sisters reproduce more successfully than those brought up in a mixed litter with unrelated siblings. What is more, female mice that mate with their biological brothers, with both sharing the genes inherited from the same mother and father, produce more offspring than females that mate with unrelated males.[40]

Some species of dinosaur were likely to have cared for their young offspring in much the same way as birds and many reptiles do. Evidence for this comes from the discovery of a fossil of the small **ornithischian dinosaur** (*Psittacosaurus* spp.), which was published in 2004. Found embedded in the rocks of the Lower Cretaceous Yixian Formation from Liaoning in China is an adult *Psittacosaurus* surrounded by 34 young similarly sized juveniles. The age of the youngsters suggests that the adult dinosaur looked after its young long after they hatched.[41]

Origin of the species

❧

Pᴇɴɢᴜɪɴꜱ ᴅɪᴅ ɴᴏᴛ evolve in freezing Antarctic climes. They first appeared in South America around 40 million years ago, when the weather was warm, and there was no ice.[1]

It is likely that some of the earliest species of **bird** actually had four wings rather than the two that we see sported by today's birds. Evidence that the evolution of flight in birds went through a 'four-winged' stage comes from a range of recently discovered fossils. The most famous is a fossil known as *Microraptor gui*, a 77-centimetre-long early bird that lived around 130 million years ago. M. *gui*, whose discovery was announced to the world in 2003, had flight feathers on all four of its limbs, and although it was not thought to have been capable of powered flight, it was believed to be a competent glider that certainly took to the air.

Further evidence for four-winged birds also comes from a fossil found in the Yixian Formation at the Jingangshan locality in Yixian, Liaoning Province, north-east China. Announced in 2004, this early bird, which lived during the Early Cretaceous period, had substantial plumage feathers attached to its upper leg, which are likely to have provided it with greater aerodynamic agility. In particular, the feathers on the legs could have helped the bird manoeuvre during flight by acting as an air brake. In the past few years, reanalysis of the most famous early bird *Archaeopteryx* has confirmed that it too had feathers on its hind limbs, although it is still unclear whether they had any aerodynamic function and actually helped *Archaeopteryx* fly.[2]

Donkeys are the only important domestic species known to have originally come from Africa. Genetic tests have revealed that this stubborn beast of burden was first bred 5,000 years ago from two African relatives, the Nubian subspecies of **wild ass** (*Equus asinus africanus*), and the **Somalian wild ass** (*Equus asinus somaliensis*). Modern cattle, sheep, goats, pigs and horses come from ancient relatives that lived outside the continent.[3]

There are 148 non-carnivorous species of mammal that routinely weigh more than 45 kilograms. Of these, however, just 14 have ever been domesticated. Thirteen of these species originated in Europe or Asia, while the **llama** is the only species to have been domesticated in the Americas.[4]

Pigs have been domesticated from wild boar on at least seven different occasions, in regions as diverse as central Italy, India, Burma, Thailand and New Guinea.[5]

All domestic breeds of poultry were bred, starting at least 8,000 years ago, from the **red jungle fowl** (*Gallus gallus*), which lived wild in south-east Asia. Compared to their wild jungle fowl cousins, modern egg-laying hens are less active, less likely to explore their environment or try new foods, interact less with other birds of the same species, and make fewer efforts to avoid predators. They also lay eggs that are roughly twice the size of those laid by jungle fowl.[6]

Of the 10,000 or so **bird** species, just ten have been domesticated.[7]

Generally speaking, small animal species have been bred to be larger to supply more meat, whereas already very large species are bred smaller to make them easier to handle. Original wild cattle, known as **aurochs**, were twice as tall as Celtic cattle bred from them. In most domestic species, head or brain size has decreased.[8]

Fish that can breathe air are thought to have evolved on at least 67 separate occasions, including many species within the family *Lepidosirenidae*, collectively known as South American lungfish, and within the family *Protopteridae* – known as African lungfish.[9]

Ants evolved from wasps.[10]

One species of **snake** has been discovered that has hips. *Najash rionegrina*, which means 'legged biblical snake' from Río Negro Province, Argentina, where the snake was discovered, lived in the Upper Cretaceous period, which lasted from 65 to 100 million years ago. It is the most primitive snake yet discovered, and uniquely among species of snake living or extinct, it has a pelvic girdle which supports robust, functional legs outside of its ribcage. The snake appears to have lived either on the surface of the land, or in subterranean holes and caves, supporting the idea that snakes evolved and lost their legs on land, rather than in the sea.[11]

No **spiders** live in Antarctica.[12]

The famous **coelacanth** (*Latimeria*), a lobe-finned primitive fish considered to be a living fossil, is more closely related to humans than it is to herrings.[13]

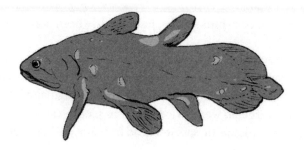

The first unique feature evolved by **tetrapods**, the group of land-living vertebrate animals that includes amphibians, reptiles, birds and mammals, was not a set of limbs. Tetrapods first began differentiating from fish when they evolved structures known as the choanae, the 'internal nostrils' that form the passage between our nasal cavity and throat, which we use for breathing when our mouth is closed.[14]

Colours of the rainbow

❧

AN ALBINO PENGUIN achieved notoriety in 2000 for being only the second white bird of its kind to find a mate and raise a family. The social stigma of being white is so strong that albinos are usually pecked at and ostracised from the penguin colony. But one plucky albino **Adélie penguin** (*Pygoscelis adeliae*) was seen by Japanese researchers incubating an egg on the edge of a colony at Amundsen Bay. One other white Adélie penguin has been seen incubating an egg in 1969, but the bird died soon after laying.[1]

The colour of a bird's egg is a good predictor of how healthy the resulting chick will be, and how likely it is that the chick will survive. The **pied flycatcher** (*Ficedula hypoleuca*) lays eggs with blue-green coloured shells, and the exact hue of the egg is related to how much the mother bird is investing in her chick. Incubating females in better condition lay more colourful eggs than those that are less healthy, because the intensity of blue-green colour reflects the amount of maternal antibodies in the yolk, a form of passive immunity imparted by the female that is crucial for the health of her offspring. Therefore chicks that hatch from brighter eggs have healthier immune systems and tend to be more successful fledglings.[2]

Very few mammals have blue, green or violet skin. The few exceptions are a handful of marsupials and primates, such as the **mandrill monkey** (*Mandrillus sphinx*), males of which not only sport a blue face, but have a vividly blue rump and scrotum. The predilection for a blue scrotum is also shared by the male **vervet monkey** (*Cercopithecus aethiops pygerythrus*), male **mouse opossum** (*Marmosa robinsoni*), and male **woolly opossum** (*Caluromys derbianusi*).[3]

Monochromatic **primates**, whose eyes can only distinguish black, white and intermediate greys, and dichromatic primates, who can see just short and middle wavelength light that includes blues and greens, tend to have furry faces. Trichromatic species, which can see blues, green and reds, tend to have bare faces, raising the idea that an ability to see changes in the skin colour of others was an important selection pressure in the evolution of colour vision in primates.[4]

Cuttlefish (*Sepia officinalis*) are renowned for their ability to change colour rapidly, almost instantaneously morphing into a range of hues that mimic the rock or seaweed of the sea floor against which they are hiding. They do so while being completely colour-blind. How cuttlefish know what disguise to adopt is therefore something of a mystery. They are able to detect differences in the contrast of their surrounding environment of at least 15 per cent, and they appear to match their body patterns to the intensities of objects nearby. But it is unclear how some cuttlefish can mimic over 30 different colour patterns when they cannot actually see the colours they are matching.[5]

A **mallard duck's** (*Anas platyrhynchos*) orange or yellow beak also glows ultraviolet light. The same is true for **blackbirds** (*Turdus merula*) and **zebra finches** (*Taeniopygia guttata*), whereas the comb of the **red grouse** (*Lagopus lagopus scoticus*) also shines in the ultraviolet spectrum.[6]

The nocturnal **helmeted gecko** (*Tarentola chazaliae*) is the only known vertebrate that can see colours in the dark. The only other animal known to have nocturnal colour vision is the **elephant hawkmoth** (*Deilephila elpenor*).[7]

Marsupials can see in colour. It was thought that primates were the only mammals that possess trichromatic vision, the ability to see colours produced at short, medium and long wavelengths of the light spectrum. However, four distantly related species of marsupial, the **fat-tailed dunnart** (*Sminthopsis crassicaudata*), **honey possum** (*Tarsipes rostratus*), a small wallaby called a **quokka** (*Setonix brachyurus*) and the bandicoot known as a **quenda** (*Isoodon obesulus*) have all been shown to possess trichromatic vision, making it likely that all marsupials have a similar ability.[8]

The albumen, or white, of all **bird** eggs is 90 per cent water.[9]

Home is where the heart is

∾

ANY INTERACTION between domestic **cats** and their owners tends to last longer if started by the cat, rather than by the owner. In families where children tend to approach the family cat more than their parents do, the cat generally prefers to interact with the adults and in particular with a woman. Cats in multi-cat households tend to rub on their owners less frequently than single cats do, and cats allowed out of doors rub more than cats confined indoors.[1]

Around 9 million domestic **cats** live in the UK, nearly 20 times the number of stoats and weasels and 38 times the number of foxes. One in three houses in the US own a cat.[2]

There is one species of especially proud **house mouse** (*Mus spretus*) to which hygiene is everything. M. *spretus* lives alongside its more common house mouse cousin M. *domesticus* across much of Europe. But M. *spretus* is unique among rodents in that it likes to clear up its own faeces. When a mouse does a fresh poo, it then either picks it up in its mouth, or rolls it along with the tip of its nose, until the droppings are discarded a long way from resting places or commonly used pathways.[3]

Peter the **penguin** became famous around the world after being rescued from an oil spill caused by a sinking tanker off Cape Town in South Africa in 2000. After being cleaned up by conservationists and released far away from the spill off the coast of Port Elizabeth, he set off to complete a 500-mile swim to his nesting grounds on Robben Island. Enthusiasts monitored his whole journey, celebrating as Peter avoided great white sharks and stormy seas. By the time of his arrival, the penguin had become a star, his survival being feted as a great success of environmental protection efforts. Media plans for a huge homecoming party were dashed, however, when Peter promptly disappeared among the thousands of other, similar-looking penguins on the island. He was not seen again for four years, until a British scientist spotted him, safe and well, in 2004.[4]

The **king penguin** (*Aptenodytes patagonicus*) and the **emperor penguin** (*Aptenodytes forsteri*) are the only two species of penguin that do not build nests. The adults incubate their eggs and warm the young chicks by tucking them between their feet and feathers.[5]

The title of most agoraphobic fish should be given to **gobies** belonging to the genus *Gobiodon*. The fish secrete poisonous mucus which makes them toxic to most predators. Despite this protection, the gobies spend their whole lives jammed in tiny crevices in coral, where no predators could possibly catch them, and have even made radical changes to their sex lives to make sure they do not have to swim outside to go looking for a mate. The fish can change sex in either direction, so if two males or females find themselves marooned together in a coral colony, one simply changes sex to form a mating pair with the other.[6]

Honeybee nests are usually paragons of organisation, with each bee setting aside its own interests to serve the queen for the good of the hive. But occasionally anarchy breaks out when all the **worker bees** (*Apis* spp.) throw off their shackles, and frantically run around the hive trying to lay their own eggs. The queen is forced to hide in a corner, protected by just a handful of loyal minions, while her subjects stage a revolution. Eventually, the workers become so disoriented that they are barely able to feed themselves, and the social order of the hive collapses completely.[7]

Hornbills take housewifery to the extreme. Female **Monteiro's hornbills** (*Tockus monteiri*) find a hole in a tree, squeeze inside, and then barricade themselves in by plugging the hole with faeces and crushed millipedes. They do this weeks before laying eggs, and spend two months in total inside their tree hollow raising their young, their only contact with the outside world being through a small gap through which they breathe, defecate, and receive food from their male partner. Their odd way of life means that female hornbills have another unique claim to fame. Once barricaded in, the male cannot reach the female to copulate, so the females mate before they enter and store the male's sperm within their bodies for up to four weeks before using it to impregnate themselves, the only birds known to do so.[8]

The European **common cuckoo** (*Cuculus canorus*) lays its eggs in the nests of over 100 other species of unsuspecting birds. However, each individual female cuckoo has a favourite species that she prefers to raise her eggs, and will preferentially lay her eggs in the nest of that species throughout her life.[9]

Killer whales form the most stable family groups of all mammals. In 25 years of intensive research watching a group of fish-eating **killer whales**, or **orcas** (*Orcinus orca*), in the coastal waters of the north-eastern Pacific Ocean, there has not been one documented incidence of a male or female offspring leaving their mother. Each baby grows up and remains within the family group for the rest of their lives.[10]

Hermit crabs often pick up new shells at the crustacean equivalent of a property market. These tend to be areas where there is a high level of crab predation on marine snails. When hermit crabs gather at such sites, a dominant crab is the first to occupy a new shell. Then the next dominant crab takes the shell made vacant by the first. This process of exchanging shells continues down the crab hierarchy until all the crabs have obtained a new, better home.[11]

Weird identity

❧

THERE IS ONE KNOWN example, confirmed by genetic tests, of a **black rhinoceros** (*Diceros bicornis*) and a **white rhinoceros** (*Ceratotherium simum*) mating and then giving birth to a black and white hybrid rhino.[1]

A female **stoat** (*Mustela erminea*) can give birth to a single litter of offspring that contains babies fathered by different males. They are able to do this because stoats carry undeveloped embryos for up to 10 months before continuing the pregnancy in the spring of the next year. Individual embryos, sired by different males, can be carried at the same time, until the female gives birth to one litter of between six and 13 babies.[2]

There is an animal whose body contains the gene of a plant. The ocean-going *Hydra viridis*, a jelly-like animal with tentacles that belongs to the *Cnidaria*, a group that also contains **jellyfish, sea anemones** and **corals**, possesses a gene called HvAPX1 that is usually found in plants, where it encodes for a chemical called ascorbate peroxidase. How this gene became part of an animal is unclear, but it is likely that *Hydra* acquired the gene much earlier in its evolutionary past, probably when it absorbed it from a symbiotic alga that lives within *Hydra*'s epithelial cells.[3]

The **platypus** (*Ornithorhynchus anatinus*) is weirder than you may think. As well as sporting a duck's bill, webbed feet, and laying eggs, despite being a furry mammal, the platypus has a very odd way of deciding whether it is male or female. Most mammals have just two sex chromosomes, the X and Y, with females inheriting two Xs and males an X and a Y. The platypus, however, has 10 such chromosomes. Females are X1X1X2X2X3X3X4X4X5X5, whereas males are X1Y1X2Y2X3Y3X4Y4X5Y5.[4]

There is a fungus that breaks a fundamental rule of biology: that all single organisms have a single genetic identity. In humans, or any other animal, each cell contains a copy of the same DNA sequence. However, the **fungus** (*Glomus etunicatum*) does things differently. Each spore contains thousands of different nuclei, each of which may carry up to 12 separate genomes, providing each organism with a multitude of genetic identities.[5]

A species of ant has taken an extreme approach to the battle of the sexes, by developing separate gene pools for males and females. The unusual reproductive system arises because queens and males, who are responsible for reproduction, individually give birth to new clones of themselves, producing entirely independent male and female generations of each sex. Only the sterile workers of the species, the **little fire ant** (*Wasmannia auropunctatai*), are produced by conventional sexual reproduction. It is also the first reported example in the animal kingdom of males reproducing clonally.[6]

A female **spotted hyena** (*Crocuta crocuta*) has a rather bizarre sexual identity. In fact, on the surface, she almost appears to be male. Female hyenas are larger than males, more aggressive and display male behaviours when playing as juveniles. They also sport highly masculine external genitalia, the most male-looking female reproductive organs of any living mammal. The female has a large, fully erectile clitoris through which she gives birth, urinates and copulates. However, when it comes to sex, female hyenas have distinctly female preferences. Although females are capable of inserting their false penises into one another, they do not attempt to copulate with each other, and will only very rarely mount one another with an erection. Even as juveniles, only males play-mount other animals.

Females become highly masculinised due to being immersed in high levels of the hormone testosterone while in their mother's womb. And that has a knock-on effect on the social status that will be granted to newborn female hyenas. Those born to mothers of high social ranking appear to be exposed to higher levels of hormone and inherit their mother's domineering personality – the first time a clear mechanism has been found by which a mammal can directly inherit its social status from its parent. It is likely that male offspring of dominant female hyenas also benefit in this way. Because the females in a pack of hyenas are so domineering, a male can find it extremely difficult to convince a female to have sex with him. Males exposed to higher levels of hormones in the womb spend more time practising their mounting behaviour when young, which gives them a higher chance of successfully mating with a female when mature.[7]

Some single-celled organisms, which spend much of their lives living free, are capable of coming together to form a single living organism. The social **amoeba** (*Dictyostelium*) usually exists as a single-cell organism that inhabits forest soil, consuming bacteria and yeast. But when starved of food, 100,000 of these individuals will come together, forming a new multicellular organism with distinct tissues, including a fruiting reproductive body made up of a stalk and spores that is capable of producing a new generation.

The **purple sponge** (*Haliclona* spp.) also releases free-swimming larvae that fuse together to form a single chimerical organism.[8]

Armadillos within the genus *Dasypus* are the only animals known where a single fertilised egg routinely divides into multiple embryos, a process known as obligate polyembryony. All offspring produced, which varies from one to 12 littermates, are genetically identical. Female **nine-banded armadillos** (*Dasypus novemcinctus*) can give birth up to three years after they have been inseminated, as the animals delay implanting the fertilised egg until the environmental conditions are right for raising young.[9]

The South American **red vizcacha rat** (*Tympanoctomys barrerae*) has the highest number of chromosomes, 51 pairs, yet described in any mammal. It also has the largest genome known of any mammal. Most mammals carry a genome of between 6 to 8 picograms of DNA in each cell's nucleus. The nuclear genome of *T. barrerae* is an extraordinary 16.8 picograms.[10]

Most animals produce offspring by fusing a sperm and an egg. But some species such as the **flatworm** (*Schmidtea polychroai*) do things a little differently. While the sperm is necessary to initiate the development of an embryo, it does not contribute any genes to the resulting offspring, a reproductive process called sperm-dependent parthenogenesis.[11]

The rarest of the rare

༄

THE STORY OF THE po'ouli, a small finch-sized honeycreeper that once adorned the remote forests of the Hawaiian island of Maui, is perhaps one of the most extraordinary, bizarre and ultimately tragic of any animal species.

Despite living on the island for hundreds, if not thousands of years, and being a striking bird with black-and-white head plumage, the **po'ouli** (*Melamprosops phaeosoma*) was unknown to native islanders until its discovery by scientists in 1973. Just nine individuals were recorded, two of which were promptly shot and killed by researchers so their bodies could be kept for future reference. By 1975, the bird was already listed as endangered.

By 1995, just six individuals were known to be living in four separate locations, and by 2000 only three birds were known to be surviving. No known breeding pairs have been discovered, while only two po'ouli nests producing one and two nestlings respectively have ever been recorded.

No one could be sure of the sex of the three survivors, but they were thought to comprise two females and a solitary male. After attempts had been made to save the bird by creating a natural reserve around its forest habitat and reducing the numbers of feral ungulates that were destroying the land, conservation biologists finally decided to catch one of the supposed females and move it into the territory of the male, in the hope they would form a breeding pair. This was done on 4 April 2002, after three weeks of searching for the bird. The day after being released, the female promptly flew back to where she was first caught, leaving the recovery effort back at square one.

On 9 September 2004, one of the three remaining po'ouli was

again captured and taken to the Maui Bird Conservation Center, in the hope of starting a captive research programme. Genetic tests on the bird's blood revealed it was a male, not a female as previously thought. The captured bird had only one functioning eye, as its right eye had completely collapsed, and was also suffering from avian malaria. After 78 days in captivity, the bird died, probably of old age.

The remaining two birds have not been seen since 2003, and the fear is that they too are both males, dashing any hopes of them mating and saving the species, or that they are nearing the end of their lives. Most tragically of all, they may both already be dead, the last survivors of their kind.[1]

Other extremely rare birds have flirted with ignominy. In 1985, a US$35 million captive breeding programme took the last remaining wild **Californian condors** (*Gymnogyps californianus*) into captivity and an intensive breeding programme was started to save the species from extinction. No chicks were born in the wild for the next 18 years. The first, in 2002, became covered in crude oil, after his father had stuck his head in a puddle near an oil well. He covered himself in oil and then returned to the chick, covering it also, but the chick did survive.[2]

The **Javan hawk eagle** (*Spizaetus bartelsi*) is a threatened raptor endemic to the densely populated island of Java, with just 600–900 birds remaining. To help save it, the bird was adopted as a national mascot, with its image appearing on billboards, postal stamps, and telephone directories. Unfortunately for the raptor, its new iconic status also appears to have attracted the attention of poachers, and the bird is now in great demand from illegal bird collectors previously unaware of its existence.[3]

Efforts to save the **Seychelles kestrel** (*Falco araea*) have been hampered by the perception of local islanders, who view the bird as an omen of bad luck. Less than 400 pairs of the bird exist.[4]

The **Alaotra grebe** (*Tachybaptus rufolavatus*) was last seen alive in Alaotra Lake in Madagscar in 1985. The remaining individuals of this flight-impaired species are thought to have been either drowned in gill nets or eaten by a predatory fish introduced into the lake upon which it lived.[5]

The world's rarest ape is not a chimp, gorilla or orang-utan, but a gibbon. The **eastern-black crested gibbon** (*Nomascus nasutus*) was once widely distributed in north-eastern Vietnam east of the Red River, in south-eastern China, and on Hainan Island. The species became extinct on the Chinese mainland in the 1950s, and in Vietnam in the 1960s, while less than 20 individuals survive on Hainan. Then, in January 2002, another 26 individuals were discovered living in the Trung Khanh district of Cao Bang Province in north-eastern Vietnam.[6]

Death becomes them ...

❧

CHIMPANZEES (*Pan troglodytes*) kill one another as often as humans living in hunter-gatherer or subsistence societies do. On average, 207 chimpanzees out of every 100,000 will be killed by one of their kind each year. Males are killed at a rate about twice those of infants and juveniles, with 977 to 1,421 males per 100,000 being killed each year compared to 476 to 736 infants and juveniles per 100,000 per year. In humans, rates of death from warfare among subsistence-society hunters and farmers are 164 and 595 per 100,000 per year, respectively.[1]

Elephants (*Loxodonta africana*) will pay particular attention to the remains of other dead elephants, spending more time investigating elephant skulls and ivory than natural objects or the skulls of other large terrestrial mammals. However, they do not pay any particular regard to the remains of relatives compared to other elephants.[2]

Apart from humans, **whales** are the only group of animals for which there is an official criterion of death. In humans, this may include the death of the whole brain, or of the brain stem. In whales, the criteria formulated to help to decide when a hunted whale is indeed dead include: a prerequisite that the hunted whale has stopped moving, defined as a relaxation of the lower jaw; or no flipper movement; or sinking without active movement.[3]

One rat, nicknamed Razza, became famous for accomplishing the rodent equivalent of the Great Escape. Razza, a single male **Norwegian rat** (*Rattus norvegicus*), was fitted with a radio-tracking collar and released all alone on the uninhabited 10-hectare New Zealand island of Motuhoropapa near Auckland. The researchers wanted to see what happens when a single rat invades an island, as a way to understand better how pests invade new territories. However, Razza gave the researchers more than they bargained for. Surrounded by traps, tempted by chocolate and even tracked by dogs, he managed to evade capture for an astonishing four months. Eventually, Razza made one final dash for freedom and leapt into the sea, swimming 400 metres across open sea to a neighbouring island, the longest documented sea crossing by a Norwegian rat. Razza's epic run-around finally came to an end when he took a rather misjudged bite out of a piece of penguin meat, which had been laid in a trap.[4]

The longest continuous satellite tracking of an individual bird came to an end on 5 March 2005, when a **white stork** (*Ciconia ciconia*) known as Donna was electrocuted when flying into power lines. First tagged in 1999, Donna was tracked for 2,033 days.[5]

Electrocution by power lines is the leading cause of death in **storks**. For instance, almost six out of ten storks that die while migrating across Israel are killed this way.[6]

At least 30 per cent of newly fledged **Spanish imperial eagles** (*Aquila adalberti*) are electrocuted each year. Fewer than 200 pairs of the raptors survive.[7]

Male nuptial **gift-giving spiders** (*Pisaura mirabilis*) will feign death to avoid being eaten by the female they are trying to mate with. Males offer gifts of food to attract a mate, but when a female approaches a gift-displaying male, she not only shows interest in the gift but will sometimes attack the male. At this moment, males often pretend to drop dead, by collapsing and remaining completely motionless. When the female has eaten the gift, the male will miraculously come back to life and attempt to mate with her.[8]

Horseracing is a dangerous sport and it is possible to determine the relative risk that a racehorse will die while running a race. Generally speaking, **horses** that run on flat races are at a lower risk of fatal injury than those running hurdle, chase or British National Hunt flat races. The longer the race, and the harder the racing surface, the greater the likelihood that a horse will be fatally injured, and novice racers are also particularly susceptible, with risk of death decreasing the more the horse has run during the previous year, or the more often it has run a particular style of race. There also appears to be a seasonal effect on how many horses die while racing, with the risks being significantly higher between February and May. Male horses are also more likely to die during a race than females. Two-year-old horses running for the first time in the UK have a fatality rate of 1.29 per 1,000 starts. Three-year-olds have a fatality rate of 2.59, while four-year-olds and older have a rate of 5.32.

TYPE OF RACE	RELATIVE HAZARD OF FATAL INJURY
Flat	1.00
National Hunt	1.46
Chase	1.47
Hurdle	1.72

GOING	RELATIVE HAZARD OF FATAL INJURY
Hard/firm	1.35
Good to firm	1.13
Good	1.00
Good to soft	0.84
Soft	0.67
Heavy	0.38

table source[9]

... or not

A **Madagascar fish eagle** (*Haliaeetus vociferoides*) managed to survive in the wild for at least seven years with only one foot. The feat is remarkable because raptors such as eagles and falcons not only need two legs to walk properly but rely on their talons to catch and carry food. The eagle, first spotted near Lake Befotaka in 1996 then again seven years later, walked on its stump, yet despite its disability it did not appear to be intimidated by other birds, holding a dominant position within the local community of fish eagles. Only two other raptors are known to have survived in the wild after losing a foot, one an immature bald eagle that survived for two years, the other a one-legged adult **Eurasian kestrel**

(*Falco tinnunculus*) that survived for one month before being struck and killed by a vehicle.[1]

In 1996 a deaf and mute dolphin was discovered by marine researchers working at US Naval Command Ocean Surveillance Center in San Diego, California. The young female **bottlenose dolphin** (*Tursiops truncatus*), aged nine years old, was monitored for another seven years, during which she never whistled or made echolocation pulses or made burst pulse sounds as other dolphins do. Despite this handicap, the dolphin was well nourished and evidently had no problems feeding or looking after herself in the wild.[2]

Fire ants (*Solenopsis invicta*) have an unusual way of surviving floods. When water levels rise, huge numbers of the ants cling together forming large rafts that float on the water's surface. By building these life rafts a colony can survive until the flood recedes or higher ground is found.[3]

Resurrection

∽

THE New Zealand petrel (*Oceanites maorianus*) came back from the dead on 25 January 2003. The species was known only from fossil remains and three specimens caught in the 19th century, and had long been considered extinct, until a birdwatcher photographed a lone individual flying off the coast of Whitianga, North Island, on that date.[1]

A male golden-crowned manakin (*Pipra vilasboasi*) was seen in Brazil in 2002. The species had not been seen alive for 45 years and was thought extinct. The same year an indigo-winged parrot (*Hapalopsittaca fuertesi*) was sighted in Colombia. It was the first time the species had been seen for 91 years.[2]

At the time of writing, at least 24 species of monkey that are new to science have been discovered since 1990. Thirteen of these new species are from Brazil including the Prince Bernhard's titi monkey (*Callicebus bernhardi*), a surprisingly conspicuous reddish-brown primate that sports a white-tipped black tail and remarkable dark orange sideburns, chest and the inner sides of its limbs, and another species of titi monkey called C. *stephennashi*, which is silver in colour, with a black forehead and red sideburns and chest.

In 2006 it was also confirmed that a new species of African monkey described for the first time the year before in Tanzania actually belonged to a new genus. The classification of the only preserved specimen of this monkey, named *Rungwecebus kipunji*, is the first time a new genus of monkey has been discovered in Africa for 83 years.[3]

You would imagine that it is hard for a whale to remain incognito for centuries. But in 2003 a new species of baleen whale was discovered, a group that contains the largest animals to have ever lived. The whale was finally identified and described for the first time from a medium-sized baleen whale carcass washed up on a coastal island in the Sea of Japan, and from eight specimens held in a museum, long thought to have been small examples of the well-known **fin whale** (*Balaenoptera physalus*). The unique morphology of the whales' skulls, DNA evidence and the number of plates of baleen the animals use to catch krill confirmed all the specimens belonged to a hitherto unknown species.[4]

In 2002, a completely new type of insect was discovered, the first time this has happened since 1914. Two museum specimens of a stick-insect-like creature, one from Tanzania, and the other from Namibia, were found to belong to a new order of insects called *Mantophasmatodea*. While new species of insect are found regularly, all belong to orders of insects, such as **beetles** (*Coleoptera*), or **dragonflies** (*Odonata*), which have been recognised for most of the 20th century. The gut contents of the museum specimens show that *Mantophasmatodea*, which grow to between 17 and 22 centimetres long, are carnivores, munching on other smaller, insect prey.[5]

Bacteria that reproduce by continually dividing into two symmetrical offspring do not live for ever. Despite appearances, the two resulting cells of the ubiquitous bug *E. coli* are not identical. One essentially becomes the parent cell, the other the juvenile offspring cell, and the older parent cell grows more slowly, produces less offspring and is more likely to die than the juvenile. In short, dividing bacteria are actually an ageing parent repeatedly producing rejuvenated offspring.[6]

A larva of the tiny **nematode worm** (*Anguillulina tritici*) can survive having all the water in its body removed and being dried to the point when the water content within its body is in chemical equilibrium with the air. When rehydrated, the larval worm regains all its bodily functions and survives the experience unscathed. This ability is known in other tiny organisms including rotifers and some species of bacteria. However, all known animals that can survive being dried out are smaller than 5 millimetres long. This may be due to the physical stresses placed on the animals' bodies, as their cells shrink as they dry, meaning the whole animal must shrink with them. This rules out any animal species that has a skeleton, and animals larger than 5 millimetres in length may be too thick to allow water to leave their bodies completely.[7]

The **dodo** (*Raphus cucullatus*) is undoubtedly dead, but it is likely to have survived for longer than people generally think. For centuries, the widely accepted date for the extinction of the dodo was 1662, when the last of the large flightless birds was believed to have been sighted on the island of Mauritius by shipwrecked mariner Volkert Evertszoon, who in that year walked on to an islet that supported a small population of the birds.

To a degree that story has been open to question due to the recollections of a man called Isaac Joan Lamotius, who became chief of the island from 1677 to 1692. Lamotius kept detailed accounts of events on the island during his tenure that included records of food collected by hunters. Within those accounts, Lamotius refers to 12 separate occasions between 1685 and 1688 when hunters collected a bird referred to as *dodaersen*. However, historians believed the word *dodaersen* was actually used to refer to another flightless bird on the island at the time, the red rail (*Aphanapteryx bonasia*).

However, new historical documents reported in 2004 brought a new perspective to the story, with the discovery of a note written by hunters reporting they had captured and killed a dodo on 16 August 1673. There are also accounts that a slave called Simon who roamed at large on the island for a short while reported to his captors that he had seen dodos on two occasions between 1663 and 1674. From these accounts and others, as well as an analysis of the rate of decline of the dodo population prior to 1662, scientists have calculated that the best guess for the actual date of the extinction of the dodo is 1690, some 28 years later than thought. There is even a 10 per cent chance that the bird finally croaked during the French occupation of Mauritius, which began in 1710.[8]

References

The mating game
1 *Animal Behaviour*, 2001, vol. 62, p. 1075
2 *Animal Behaviour*, 2006, vol. 71, p. 315
3 *Animal Behaviour*, 2003, vol. 66, p. 175
4 *Auk*, 2003, vol. 120, p. 717
5 *Science*, 2006, vol. 311, p. 965
6 *Science*, 2004, vol. 304, p. 5670
7 *Animal Behaviour*, 2006, DOI:10.1016/j.anbehav.2005.06.023
8 *Proceedings of the Royal Society B*, 2005, vol. 272, p. 2485
9 *Proceedings of the Royal Society B*, 2005, vol. 272, p. 319
10 *Nature*, 2004, vol. 432, p. 1024
11 *Proceedings of the Royal Society B*, 1999, DOI:10.1098/rspb.1999.0761
12 *Behavioral Ecology and Sociobiology*, 2004, DOI:10.1007/s00265-004-0857-7
13 *Journal of Experimental Biology*, 2005, vol. 208, p. 3433
14 Ibid.
15 *Journal of Ethology*, 2005, vol. 23, p. 51
16 *Behavioral Ecology*, 2006, vol. 16, p. 377
17 *Animal Behaviour*, 2006, vol. 71, p. 481
18 *Proceedings of the National Academy of Sciences*, 2004, vol. 101, p. 4883
19 *Animal Behaviour*, 2006, vol. 71, p. 481
20 *Animal Behaviour*, 2006, DOI:10.1016/j.anbehav.2005.06.023
21 *Animal Behaviour*, 2006, vol. 71, p. 481
22 *Animal Behaviour*, 2001, vol. 61, p. 497
23 *Nature*, 2004, vol. 429, p. 551
24 The Society for Integrative and Comparative Biology annual meeting 2003, Toronto
25 *Nature*, 2004, vol. 431, p. 446
26 *International Journal of Primatology*, 2004, vol. 25, p. 1159
27 The Society for Integrative and Comparative Biology annual meeting 2003, Toronto
28 *Proceedings of the Royal Society B*, 2005, vol. 272, p. 365
29 *Behavioral Ecology and Sociobiology*, 2005, vol. 57, p. 457
30 *Animal Behaviour*, 2000, vol. 59, p. 349
31 *Animal Behaviour*, 2004, vol. 67, p. 1043
32 *Journal of Experimental Biology*, vol. 208, p. 647
33 *Journal of Environmental Radioactivity*, 2005, vol. 79, p. 1
34 *Animal Behaviour*, 2005, vol. 69, p. 529
35 *Animal Behaviour*, 2005, vol. 69, p. 529
36 *Experimental and Applied Acarology*, 2005, vol. 35, p. 173
37 *Journal of Experimental Biology*, 2005, vol. 208, p. 2037
38 *Hormones and Behavior*, 2006, DOI:10.1016/j.yhbeh.2006.01.001
39 *Animal Reproduction Science*, 2005, vol. 86, p. 353

Unhealthy living

1 *Journal of Exotic Pet Medicine*, 2006, vol. 15, p. 59
2 *Current Biology*, 2004, vol. 14, p. R988
3 Ibid.
4 *Journal of Antimicrobial Chemotherapy*, 2006, vol. 57, p. 461
5 *Environmental Biology of Fishes*, 2005, vol. 72, p. 155
6 *Aerobiologia*, 2005, vol. 21, p. 147
7 *Biological Review*, 1996, vol. 71, p. 415
8 *Parasitology Today*, 1992, vol. 8, p. 159
9 *Parasitology Today*, 1992, vol. 8, p. 159
10 David Macdonald (ed.), *The New Encyclopedia of Mammals*, Oxford University Press, 2001, p. 796
11 *Nature*, 2006, vol. 441, p. 421

Drink and drugs

1 *New Scientist*, 2003, vol. 180, p. 56
2 *Integrative and Comparative Biology*, 2004, vol. 44, p. 295
3 *Primates*, 2004, vol. 45, p. 113
4 *Animal Behaviour*, 2006, DOI:10.1016/j.anbehav.2005.09.012
5 *New Scientist*, 2003, vol. 180, p. 56
6 Ibid.
7 Ibid.
8 *Proceedings of the National Academy of Sciences*, 2005, 10.1073/pnas.0406814102
9 *Trends in Pharmacological Sciences*, 2005, vol. 26, p. 287
10 *Journal of Archaeological Science*, 2006, vol. 33, p. 158

Talented creatures

1 *Animal Cognition*, 2004, vol. 7, p. 231
2 *Animal Cognition*, 2003, DOI:10.1007/s10071-003-0191-x
3 David Macdonald (ed.), *The New Encyclopaedia of Mammals*, Oxford University Press, 2001, p. 17
4 *Animal Cognition*, 2003, DOI:10.1007/s10071-003-0202-y
5 *Animal Cognition*, 2004, vol. 7, p. 69
6 *Nature*, 2004, vol. 431, p. 39
7 *Nature*, 2004, vol. 431, p. 39
8 *Biology Letters*, 2005, DOI:10.1098/rsbl.2004.0291
9 *Science*, 2006, vol. 311, p. 1297
10 Macdonald (ed.), *The New Encyclopaedia of Mammals*, 2001, p. 114
11 *Journal of Experimental Psychology: Applied Behavior Processes*, 2004, vol. 31, p. 95
12 *Nature*, 2002, DOI:10.1038/news020603-12
13 *Animal Behaviour*, 2005, vol. 69, p. 209
14 *Current Biology*, 2006, vol. 16, p. 512
15 *Journal of Experimental Biology*, 2005, vol. 208, p. 4709

16 *Applied Animal Behaviour Science*, 2003, vol. 81, p. 307
17 *Current Biology*, 2005, vol. 15, p. 543
18 *Primates*, 1995, vol. 36, p. 259
19 *Journal of Ethology*, 2005, DOI:10.1007/s10164-004-0132-4
20 *Journal of Ethology*, 2005, vol. 23, p. 9
21 *Journal of Arid Environments*, 1996, vol. 32, p. 453
22 *Nature*, 2004, vol. 430, p. 417

Talking ...

1 *Journal of Comparative Psychology*, 1993, vol. 107, p. 301
2 *Journal of the Acoustical Society*, 2006, vol. 119, p. 627
3 *Journal of the Acoustical Society of America*, 2006, DOI:10.1121/1.2161827
4 *Nature*, 2004, vol. 430, p. 523
5 *PLoS Biology*, 2005, vol. 3, issue 12
6 *International Journal of Primatology*, 2005, vol. 26, p. 73
7 *Journal of Comparative Physiology A*, 2004, vol. 190, p. 791
8 *The Journal of the Acoustical Society of America*, 2004, vol. 116, p. 2640
9 *Animal Behaviour*, 2004, DOI:10.1016/j.anbehav.2003.07.016
10 *Mammal Review*, 2002, vol. 32, p. 245
11 Ibid.
12 American Neurological Association meeting, Seattle, October 1999
13 *Behavioral Ecology and Sociobiology*, 2003, vol. 55, p. 415
14 *Behavioral Ecology and Sociobiology*, 2006, DOI:10.1007/s00265-006-0188-y
15 *Science*, 2005, vol. 308, p. 1934
16 *Nature*, 2004, vol. 431, p. 146
17 *Animal Behaviour*, 2004, vol. 67, p. 855

and listening ...

1 *Journal of Experimental Biology*, 2004, vol. 207, p. 427
2 *Journal of Experimental Biology*, 2005, vol. 208, p. 3533
3 Ibid.
4 *Behavioural Processes*, 1998, vol. 43, p. 211
5 *Science*, 2005, vol. 307, p. 910
6 *Journal of Experimental Biology*, 2004, vol. 207, p. 155
7 *Nature*, 2006, vol. 440, p. 333

Eating ...

1 *New Scientist*, 24 November 1990, vol. 128, p. 48
2 *PLoS Genetics*, 2005, DOI:10.1371/journal.pgen.0010003
3 *Biology Letters*, 2006, DOI:10.1098/rsbl.2006.0480
4 *Nature*, 2006, vol. 440, p. 930
5 *Animal Behaviour*, 2006, DOI:10.1016/j.anbehav.2005.10.007
6 *Animal Behaviour*, 2006, DOI:10.1016/j.anbehav.2005.07.022
7 California Academy of Sciences

8 *Animal Behaviour*, 2003, vol. 65, p. 385

9 *Polar Biology*, 2003, vol. 26, p. 115

10 *Animal Behaviour*, 2004, vol. 68, p. 583

11 *Nature*, 2004, vol. 433, p. 519

12 *Nature*, 2004, vol. 430, p. 772

13 *Nature*, 2002, vol. 418, p. 143

14 *Nature*, 2005, vol. 434, p. 37

15 *Journal of Experimental Biology*, 2004, vol. 207, p. 2351

16 *Naturwissenschaften*, 2006, vol. 93, p. 145

17 *Toxicon*, 2004, vol. 43, p. 471

18 *Behavioral Ecology and Sociobiology*; 2005, vol. 57, p. 267

19 *Oecologia*, 2003, vol. 136, p. 14

20 *Journal of Ethology*, 2005, vol. 23, p. 19

21 *Journal of Experimental Biology*, 2005, vol. 208, p. 1277

22 *Applied Animal Behaviour Science*, 2006, vol. 97, p. 17

23 *Journal of Theoretical Biology*, 1993, vol. 160, p. 427

24 Ibid.

25 *Mammal Review*, 2005, vol. 35, p. 123

26 *Journal of Zoology*, 2005, vol. 267, p. 309

27 *Journal of Human Evolution*, 2006, vol. 50, p. 377

28 *Journal of Archaeological Science*, 2003, DOI:10.1016/j.jas.2003.12.005

29 *Behavioral Ecology and Sociobiology*, 2006, DOI:10.1007/s00265-006-0166-4

30 *Aquarium Sciences and Conservation*, 2001, vol. 3, p. 225

31 *Nature*, 2004, vol. 432, p. 969

32 *Journal of Zoology*, 2006, vol. 269, p. 29

33 *New Scientist*, 3 September 2005, p. 85

34 *Physiology & Behavior*, 2006, vol. 87, p. 255

and excreting...

1 *Polar Research*, 2003, vol. 27, p. 56

2 *New Scientist*, 2003, vol. 177, p. 11

3 *Journal of Sea Research*, 2006, DOI:10.1016/j.seares.2006.03.003

4 *Journal of Archaeological Science*, 2004, DOI:10.1016/j.jas.2004.02.004

5 *Vaccine*, 2004, vol. 22, p. 3976

6 Ibid.

7 *Journal of Experimental Biology*, 2005, vol. 208, p. 3281

Species that regularly eat their own faeces

1 *Mammal Review*, 2001, vol. 31, p. 61

Breathe easy

1 *Journal of Experimental Biology*, 2004, vol. 207, p. 973

2 *Journal of Experimental Biology*, 2005, vol. 208, p. 3645

3 *Nature*, 2005, vol. 433, p. 516

4 *Journal of Experimental Biology*, 2005, vol. 208, p. 1373

Under pressure
1 *Animal Behaviour*, 2003, vol. 66, p. 1085
2 *Nature Neuroscience*, 2005, DOI:10.1038/nn1399
3 *Behavioural Brain Research*, 2006, vol. 168, p. 47
4 *The Veterinary Record*, 1995, vol. 136, p. 431
5 *Biology Letters*, 2005, DOI:10.1098/rsbl.2004.0245
6 David Macdonald (ed.), *The New Encyclopaedia of Mammals*, Oxford University Press, 2001, p. 246
7 *Applied Animal Behaviour Science*, 2001, vol. 73, p. 93
8 *Nature*, 1998, DOI:10.1038/news981105-1
9 *PLoS Biology*, 2004, DOI:10.1371/journal.pbio.0020212
10 *Science*, 2004, DOI:10.1126/science.1105452
11 *Toxicon*, 2004, vol. 43, p. 471
12 *Biology Letters*, 2004, DOI:10.1098/rsbl.2004.0250
13 *Journal of Experimental Biology*, 2006, vol. 209, p. 702

Good vibrations
1 *Journal of Experimental Biology*, 2005, vol. 208, p. 3103
2 *Naturewissenshaften*, 1999, vol. 86, p. 544
3 *New Scientist*, 2000, vol. 166, p. 38
4 *Journal of Experimental Biology*, 2005, vol. 208, p. 647
5 *Journal of Experimental Biology*, 2005, vol. 208, p. 3421

Questions, questions, questions ...
1 *Mammal Review*, 2002, vol. 32, p. 237
2 *New Scientist*, 2000, vol. 165, p. 22
3 Ibid.
4 *Journal of Experimental Biology*, 2004, vol. 207, p. 607
5 *Animal Behaviour*, 2003, vol. 66, p. 893 and 2005, vol. 70, p. 1265
6 *Behavioral Ecology and Sociobiology*, 2003, 10.1007/s00265-003-0726-9
7 *Animal Behaviour*, 2004, vol. 67, p. 379
8 University at Buffalo, the State University of New York
9 The Society for Integrative and Comparative Biology annual meeting 2003, Toronto
10 *Journal of Zoology*, 2006, vol. 268, p. 415
11 *Journal of Mammalogy*, 2006, vol. 87, p. 345
12 *Zoological Journal of the Linnean Society*, 1998, vol. 123, p. 1
13 *Journal of Zoology*, 2006, vol. 269, p. 89

Enduring enigmas
1 The Society for Integrative and Comparative Biology annual meeting 2003, Toronto; Science, 2002, vol. 297, p. 1339; Animal Behaviour, 2006, vol. 71, p. 609; Journal of Mammology, 2006, vol. 87, p. 193
2 *Integrative and Comparative Biology*, 2004, vol. 44, p. 203
3 Society of Vertebrate Paleontology meeting, Denver, 2004

4 *Journal of Human Evolution*, 2006, vol. 50, p. 414

5 *European Journal of Wildlife Research*, 2005, DOI:10.1007/s10344-005-0093-0

6 *Proceedings: Biological Sciences*, 2005, DOI:10.1098/rspb.2004.2984

Mysterious giants

1 *New Scientist*, 9 October 2004, p. 32 and XXI Congress of the International Primatological Society 2006

2 *New Scientist*, 15 June 2002, p. 35

3 *The Journal of the Acoustical Society of America*, 2005, vol. 118, p. 3346

4 *The Biological Bulletin*, 2004, vol. 206, p. 125

5 *Deep-Sea Research*, 2004, vol. 51, p. 17 and *New Scientist*, 02 August 2003, p. 24 and 20 December 2003, p. 23

6 *Deep-Sea Research*, 2004, vol. 51, p. 1889

7 *Hereditas*, 1991, vol. 115, p. 183

Man and beast don't get along

1 *Biological Conservation*, 2004, vol. 119, p. 279

2 *Biological Conservation*, 2005, vol. 126, p. 339

3 *Biological Conservation*, 2004, vol. 118, p. 449

4 *Biological Conservation*, 2005, DOI:10.1016/j.biocon.2005.01.007

5 *Naturwissenschaften*, 2005, DOI:10.1007/s00114-004-0606-9

6 *Biological Conservation*, 2006, vol. 130, p. 530

7 *Animal Behaviour*, 1999, vol. 57, p. 741

8 *Proceedings of the Royal Society London B*, 2004 (suppl.), DOI:10.1098/rsbl.2004.0201

9 *The Journal of Experimental Biology*, 2004, vol. 207, p. 2694

10 *Animal Behaviour*, 2004, vol. 67, p. 1163

11 *Animal Behaviour*, 2004, vol. 68, p. 649 and p. 665

12 *Journal of Emergency Nursing*, 2004, vol. 30, p. 542

13 *American Journal of Preventive Medicine*, 2001, vol. 21, p. 209

14 *Journal of Safety Research*, 1999, vol. 30, p. 219

15 *Biological Conservation*, 2006, DOI:10.1016/j.biocon.2006.01.001

16 *Animal Behaviour*, 2006, vol. 71, p. 389

17 *European Journal of Wildlife Research*, 2005, DOI:10.1007./s10344-005-0006-2

18 *Biological Conservation*, 2005, vol. 121, p. 419

19 *Animal Behaviour*, 2006, vol. 71, p. 491

20 *New Scientist*, 22 April 2006, p. 11

21 *New Scientist*, 29 October 2005, p. 34

22 David Macdonald (ed.), *The New Encyclopedia of Mammals*, Oxford University Press, 2001, p. 249

Differences

1 *Mammal Review*, 2005, vol. 35, p. 215

When animals attack
1 International Shark Attack File 2006, Florida Museum of Natural History
2 Australian Fisheries Museum Webite: www.amonline.net.au
3 International Shark Attack File 2006, Florida Museum of Natural History
4 World Heritage Tropical Forests Conference, Cairns, 1996
5 David Macdonald (ed.), *The New Encyclopaedia of Mammals*, Oxford University Press, 2001, p. 71
6 Ibid., p. 20
7 *Tropical Medicine & International Health*, 2000, vol. 5, p. 507
8 *Toxicon*, 2004, vol. 43, p. 471
9 Ibid., p. 477
10 *Toxicon*, 1996, vol. 34, p. 324
11 *New Scientist* online, 3 November 2005
12 *Nature*, 2005, DOI:10.1038/nature04328
13 *New Scientist*, 24 January 2004, p. 15

Nature red in tooth and claw
1 *Journal of Experimental Biology*, 2006, vol. 209, p. 1413
2 *Animal Behaviour*, 2006, DOI:10.1016/j.anbehav.2005.06.023
3 *Nature*, 2006, vol. 440, p. 881
4 *Conservation Biology*, 2003, vol. 17, p. 1738
5 *Oecologia*, 2004, DOI:10.1007/s00442-004-1609-0
6 *Biology Letters*, 2005, DOI:10.1098/rsbl.2004.0242
7 *Animal Behaviour*, 2005, vol. 70, p. 59
8 *Journal of Experimental Biology*, 2006, vol. 209, p. 1430
9 *Nature*, 2004, vol. 432, p. 283
10 *Behavioral Ecology and Sociobiology*, 2005, DOI:10.1007/s00265-004-0851-0
11 *Nature*, 2005, vol. 433, p. 624
12 *Journal of Ethology*, 2004, DOI:10.1007/s10164-004-0119-1
13 *Behavioural Processes*, 2005, vol. 68, p. 145
14 *Primates*, 2006, 10.1007/s10329-005-0168-2
15 *Animal Behaviour* 2004, vol. 67, p. 59
16 *Journal of Experimental Marine Biology and Ecology*, 2006, vol. 329, p. 206
17 *Toxicon*, 2002, vol. 40, p. 749
18 *Naturwissenschaften*, 2006, DOI:10.1007/s00114-006-0094-1
19 *Behavioral Ecology and Sociobiology*, 2003, vol. 55, p. 32
20 *Journal of Zoology*, 2006, DOI:10.1111/j.1469-7998.2006.00079.x
21 *Biology Letters*, 2005, DOI:10.1098/rsbl.2004.0237
22 *Nature*, 2004, vol. 430, p. 975
23 *Journal of Experimental Biology*, 2005, vol. 208, p. 3475
24 *Applied Animal Behaviour Science*, 2006, vol. 96, p. 83
25 *Mammal Review*, 2003, vol. 33, p. 174
26 *Science*, 2004, vol. 304, p. 5670
27 *Proceedings of the National Academy of Sciences*, 2006, DOI:10.1073/pnas.0508915103

28 *Nature*, 2006, vol. 440, p. 756
29 *Toxicon*, 1996, vol. 34, p. 293
30 *Journal of Zoology*, 2006, vol. 268, p. 361
31 *Proceedings of the Royal Society B*, 2005, DOI:10.1098/rspb.2004.2986
32 *Proceedings of the Royal Society B*, 2006, DOI:10.1098/rspb.2006.3532
33 *International Congress Series*, 2004, vol. 1275, p. 351
34 *European Journal of Wildlife Research*, 2005, DOI:10.1007/s10344-005-0102-3
35 *Biologist*, 2005, vol. 52, p. 339
36 *Proceedings of the Royal Society B*, 2006, DOI:10.1098/rspb.2006.3532

Biomechanical wonders

 1 The Society for Integrative and Comparative Biology annual meeting 2003, Toronto
 2 *Current Biology*, 2006, vol. 16, p. 767
 3 *Proceedings of the Royal Society B*, 2004, DOI:10.1098/rspb.2003.2637
 4 *Journal of Experimental Biology*, 2001, vol. 204, p. 3621
 5 *Journal of Experimental Biology*, 2004, vol. 207, p. 21
 6 *Journal of Experimental Biology*, 2004, vol. 207, p. 1361
 7 *Journal of Experimental Biology*, 2005, vol. 208, p. 2147
 8 *Journal of Experimental Biology*, 2005, vol. 208, p. 3293
 9 *Journal of Experimental Biology*, 2005, vol. 208, p. 3409
10 *Proceedings of the Royal Society A: Mathematical, Physical and Engineering Sciences*,
    2006, DOI:10.1098/rspa.2006.1718
11 *Solar Energy Materials and Solar Cells*, 2006, vol. 90, p. 1458
12 M.B Thompson and B.K. Speake, 'Egg morphology and composition', in D.C.
    Deeming (ed.), *Reptilian Incubation: Environment, Evolution and Behaviour*,
    Nottingham University Press, 2004, pp 45–74
13 *Journal of Experimental Biology*, 2006, vol. 209, p. 141
14 *Journal of Experimental Biology*, 2005, vol. 208, p. 3655
15 *Science*, 2005, vol. 309, p. 275
16 *Science*, 2005, vol. 309, p. 253
17 *Proceedings of the National Academy of Sciences*, 2006, vol. 103, p. 5764
18 *Science*, 1999, vol. 284, p. 798
19 *Science*, 2003, vol. 302, p. 549
20 *Journal of Experimental Biology*, 2004, vol. 207, p. 1577
21 *Journal of Zoology*, 1999, vol. 247, p. 257
22 David Macdonald (ed.), *The New Encyclopaedia of Mammals*, Oxford University
    Press, 2001, p. 800
23 *Journal of Experimental Biology*, 2005, vol. 208, p. 2913
24 The Society for Integrative and Comparative Biology annual meeting 2003, Toronto
25 *Physiological and Biochemical Zoology*, 2006, vol. 79, p. 20
26 *Nature*, 2005, vol. 434, p. 383
27 Macdonald (ed.),*The New Encyclopaedia, of Mammals*, 2001, p. 9
28 *Proceedings of the Royal Society B*, 2006, DOI:10.1098/rspb.2006.3533
29 *Journal of Comparative Physiology B*, 2003, vol. 173, p. 583

30 *Journal of Biomechanics*, 2004, vol. 37, p. 1513

31 *Science*, 2005, vol. 305, p. 1396

32 *Toxicon*, 2004, vol. 43, p. 471

33 Ibid.

34 *Journal of Experimental Biology*, 2006, vol. 209, p. 1463

35 *Journal of Experimental Biology*, 2005, vol. 208, p. 2633

36 *Biology Letters*, 2005, DOI:10.1098/rsbl.2004.0269

37 *Proceedings of the National Academy of Sciences*, 2004, vol. 101, p. 15857

38 *European Journal of Wildlife Research*, 2004, DOI:10.1007/s10344-004-0058-8

39 *Nature*, 2004, vol. 429, p. 363

40 Physiological ecology laboratory website, Princeton University, 2006, www.princeton.edu/~wikelski/Research.htm

41 *Integrative and Comparative Biology*, 2003, vol. 43, p. 376

Super organs

1 *Behavioural Ecology and Sociobiology*, 1999, vol. 46, p. 267

2 *Proceedings of the National Academy of Sciences*, 2006, vol. 103, p. 4381

3 *Animal Cognition*, 2004, DOI:10.1007/s10071-004-0219-x

4 The Society for Integrative and Comparative Biology annual meeting 2003, Toronto

5 *Science*, 2006, vol. 311, p. 1617

6 *Journal of Experimental Biology*, 2006, vol. 209, p. 18

7 The Society for Integrative and Comparative Biology annual meeting 2003, Toronto

8 *Current Biology*, 2005, vol. 15, p. R626

9 *Oikos*, 2004, vol. 105, p. 255

10 *Proceedings of the Royal Society B: Biological Sciences*, 2005, DOI:10.1098/rspb.2005.3367

11 *Current Biology*, 2003, vol. 13, p. 493

12 *Journal of Experimental Biology*, 2004, vol. 207, p. 553

13 David Macdonald (ed.), *The New Encyclopaedia of Mammals*, Oxford University Press, 2001, p. 800

14 *Animal Reproduction Science*, 2005, vol. 86, p. 205

15 *Proceedings of the Royal Society B*, 2005, vol. 272, p. 149

16 *The Journal of Experimental Biology*, 1999, vol. 202, p. 2719

17 *Philosophical Transactions of the Royal Society of London B*, 2004, vol. 359, p. 765

18 *Journal of Evolutionary Biology*, 2005, DOI:10.1111/j.1420-9101.2004.00805.x

19 *Sciencexpress*, 2006, DOI:10.1126/science.1123497

20 *Current Biology*, 2004, vol. 14, p. R557

21 Cornell University website, www.news.cornell.edu/stories/March06/MouseSensor.kr.html

Energy efficient

1 *Nature*, 2004, vol. 429, p. 825

2 *Journal of Experimental Biology*, 2005, vol. 208, p. 1373

3 *Naturwissenschaften*, 2006, vol. 93, p. 80

4 *Journal of Zoology*, 2006, vol. 269, p. 65
5 American Association for the Advancement of Science annual meeting, Denver, Colorado, 2003
6 *Oikos*, 2004, vol. 107, p. 583
7 *Journal of Experimental Biology*, 2005, vol. 208, p. 707
8 *Journal of Experimental Biology*, 2006, vol. 209, p. 1404
9 *Journal of Experimental Biology*, 2006, vol. 209, p. 1421
10 *Journal of Experimental Biology*, 2005, vol. 208, p. 771
11 *Journal of Experimental Biology*, 2004, vol. 207, p. 667
12 *Small Ruminant Research*, 2005, vol. 60, p. 111
13 *Journal of Experimental Biology*, 2005, vol. 208, p. 2595
14 *Journal of Experimental Biology*, 2005, vol. 208, p. 2533
15 *Journal of Experimental Biology*, 2005, vol. 208, p. 2549
16 *Current Biology*, 2004, vol. 15, p. 64
17 *Journal of Experimental Biology*, 2004, vol. 207, p. 4383
18 *Journal of Thermal Biology*, 2001, vol. 26, p. 555
19 *Journal of Experimental Biology*, 2004, vol. 207, p. 391
20 *Behavioral Ecology and Sociobiology*, 2004, DOI:10.1007/s00265-004-0906-2
21 *Journal of Experimental Biology*, 2004, vol. 207, p. 579
22 *Physiology Biochemistry Zoology*, 2003, vol. 76, p. 716
23 *Animal Behaviour*, 2006, vol. 71, p. 523
24 Ibid.
25 *Primates*, 2006, DOI:10.1007/s10329-005-0171-7
26 *Biology Letters*, 2005, DOI:10.1098/rsbl.2004.0243
27 *Nature*, 2006, vol. 440, p. 795
28 *Journal of Experimental Biology*, 2005, vol. 208, p. 1469

Do the locomotion

1 *Nature*, 2003, vol. 425, p. 707
2 *Nature*, 2005, vol. 438, p. 753
3 *Nature*, 2003, DOI:10.1038/426785a
4 Ibid.
5 *Journal of Experimental Biology*, 2004, vol. 207, p. 713
6 *Journal of Experimental Biology*, 2004, vol. 207, p. 461
7 *Journal of Experimental Biology*, 2004, vol. 207, p. 3507
8 *Journal of Experimental Biology*, 2004, vol. 207, p. 3545
9 The Society for Integrative and Comparative Biology annual meeting 2003, Toronto
10 *Journal of Experimental Biology*, 2005, vol. 208, p. 2741
11 *Science*, 2005, vol. 307, p. 1927
12 *Journal of Experimental Biology*, 2004, vol. 207, p. 3693
13 *Journal of Experimental Biology*, 2006, vol. 209, p. 1617
14 *Journal of Experimental Biology*, 2005, vol. 208, p. 1489
15 *Journal of Experimental Biology*, 2004, vol. 207, p. 697
16 *Physics of Fluids*, 2004, vol. 16, p. L39

17 *Journal of Experimental Biology*, 2006, vol. 209, p. 590

18 *Science*, 2005, vol. 310, p. 100

19 *Animal Behaviour*, 2004, DOI:10.1016/j.anbehav.2003.11.007

20 The Society for Integrative and Comparative Biology annual meeting 2004, New Orleans

21 *Journal of Experimental Biology*, 2005, vol. 208, p. 1435

22 *Nature*, 2005, vol. 434, p. 292

23 *Journal of Experimental Biology*, 2005, vol. 208, p. 1309

24 *International Journal of Primatology*, 2005, vol. 26, p. 223

25 *Zoology*, 2006, vol. 109, p. 43

26 *Journal of Human Evolution*, 2006, vol. 50, p. 469

27 Ibid.

28 *Zoology*, 2005, vol. 108, p. 107

29 *Current Biology*, 2005, vol. 15, p. R243

30 *Proceedings of the Royal Society B*, 2006, DOI:10.1098/rspb.2006.3532

31 *Journal of Experimental Biology*, 2005, vol. 208, p. 1817

32 The Society for Integrative and Comparative Biology annual meeting 2004, New Orleans

33 *Nature*, 1997, vol. 386, p. 269

Left or right

1 *Behavioural Ecology*, 2000, vol. 11, p. 411

2 *BMC Ecology*, 2003, vol. 3, p. 9

3 *Applied Animal Behaviour Science*, DOI:10.1016/j.applanim.2004.11.001

4 *Animal Behaviour*, 2002, vol. 64, p. 461 and 2004, vol. 68, 1107

5 *International Journal of Primatology*, 2004, vol. 25, p. 1243

6 Ibid.

In the blood

1 *Transfusion Medicine Reviews*, vol. 18, 2004, p. 117

2 Ibid.

3 Ibid.

4 Ibid.

5 Ibid.

6 *Biologist*, 2005, vol. 52, p. 349

Smelly world

1 *Nature*, 2005, vol. 438, p. 1097

2 *Current Biology*, 2005, vol. 15, p. R255

3 *Neuroscience & Biobehavioral Reviews*, 2001, vol. 25, p. 597

4 *Science*, 2004, vol. 36, p. 835

5 *Journal of Experimental Biology*, 2005, vol. 208, p. 4199

6 *Behavioral Ecology and Sociobiology*, 1999, vol. 47, p. 29

7 *Science*, 2006, vol. 311, p. 666

8 *Biology Letters*, 2006, DOI:10.1098/rsbl.2006.0448
9 *New Scientist*, 22 April 2006, p. 42
10 Ibid.

Zoological definitions
1 David Macdonald (ed.), *The New Encyclopaedia of Mammals*, Oxford University Press, 2001, and *Current Biology*, 2005, vol. 15, p. R255

The power of the goat
1 *Small Ruminant Research*, 2005, vol. 60, p. 13

Ripe for mummification
1 *New Scientist* online, 8 February 2005 and 18 September 2004

Laugh and play
1 *Primates*, 2004, vol. 45, p. 221
2 *Consciousness and Cognition*, 2005, vol. 14, p. 30
3 *Physiology & Behavior*, 2003, vol. 79, p. 533
4 *Behavioural Brain Research*, 2002, vol. 134, p. 31

Amazing fakes
1 *Molecular Phylogenetics and Evolution*, 2002, vol. 23, p. 91
2 *New Scientist*, 1993, vol. 140, p. 4
3 *Nature*, 2001, vol. 408, p. 705
4 *Biology Letters*, 2005, DOI:10.1098/rsbl.2004.0272

Mistaken identity
1 University of Michigan, Museum of Zoology website, animaldiversity.ummz.umich.edu/site/accounts/information/Dermoptera.html
2 David Macdonald (ed.), *The New Encyclopaedia of Mammals*, Oxford University Press, 2001, p. 95 and *Molecular Phylogenetics and Evolution*, 2001, vol. 17, p. 190
3 Macdonald (ed.), *The New Encyclopaedia*, p. 221
4 Macdonald (ed.), *The New Encyclopaedia*, p. 793
5 The Society for Integrative and Comparative Biology annual meeting 2004, New Orleans
6 Macdonald (ed.), *The New Encyclopaedia*, p. 93
7 *Placenta*, 2001, vol. 22, p. 800
8 Macdonald (ed.), *The New Encyclopaedia*, p. 47
9 *Wildlife Research*, 2003, vol. 30, p. 213
10 *European Journal of Wildlife Research*, 2005, DOI:10.1007/s10344-005-0093-0
11 *Proceedings of the National Academy of Science*, 2004, DOI:10.1073/pnas.0306243101
12 *Animal Behaviour*, 2004, vol. 68, p. 583
13 *Proceedings of the Royal Society B*, 2006, DOI:10.1098/rspb.2006.3500
14 *Proceedings of the National Academy of Sciences*, 2004, vol. 101, p. 15857

15 *Biology Letters*, 2004, vol. 271, p. 416
16 *Nature*, 2004, vol. 430, p. 309
17 *Systematics and Biodiversity*, 2005, vol. 2, p. 419 and *Science*, 2006, vol. 311, p. 1456

Telling look-alikes apart
1 David Macdonald (ed.), *The New Encyclopaedia of Mammals*, Oxford University Press, 2001, p. 147
2 Crocodilian Biology Database, Florida Museum of Natural History
3 Macdonald (ed.), *The New Encyclopaedia*, p. 239
4 The Field Museum, Chicago

Alliterative collective nouns
1 San Diego Zoo.org

What's in a name?
1 *Environmental Biology of Fishes*, 2002, vol. 65, p. 249
2 David Macdonald (ed.), *The New Encyclopaedia of Mammals*, 2001, p. 176

More than one name to call a cat
1 Mark Carwardine, *The Guinness Book of Animal Records*, Guinness Publishing Ltd, 1995, and IUCN Cat Specialist Group

That time of life
1 *Primates*, 2003, vol. 45, p. 73
2 *Journal of Human Evolution*, 2004, vol. 47, p. 385
3 *Biology Letters*, 2006, DOI:10.1098/rsbl.2005.0426
4 *International Journal of Primatology*, 2005, vol. 26, p. 3
5 *Science*, 2006, DOI:10.1126/science.1122446
6 *Science*, 2006, vol. 311, p. 1301
7 BIO 2005 conference, Philadelphia and *Biochemical and Biophysical Research Communications*, 1989, vol. 164, p. 1380 and 1988, vol. 154, p. 529
8 *Nature*, 2005, vol. 435, p. 1177
9 *Proceedings of the National Academy of Sciences*, 2005, vol. 102, p. 14860
10 *Nature*, 2004, vol. 431, p. 838
11 *Journal of Experimental Biology*, 2004, vol. 207, p. 1113
12 *Evolutionary Ecology*, 2006, vol. 20, p. 143
13 *Nature*, 2006, vol. 440, p. 926
14 *Nature*, 2005, vol. 438, p. 929
15 *Biological Conservation*, 2006, vol. 130, p. 604
16 *Science*, 1997, vol. 153, p. 20
17 *Biological Conservation*, 2006, vol. 130, p. 604
18 *Current Biology*, 2005, vol. 15, p. R288
19 *Journal of Experimental Biology*, 2004, vol. 207, p. 2133
20 The Society for Integrative and Comparative Biology annual meeting 2003, Toronto

21 David Macdonald (ed.), *The New Encyclopaedia of Mammals*, Oxford University Press, 2001, p. 202
22 *Science*, 2006, vol. 310, p. 630
23 The Society for Integrative and Comparative Biology annual meeting 2003
24 *Proceedings of the National Academy of Sciences*, 2006, vol. 103, p. 6587
25 *Journal of Arid Environments*, 1996, vol. 32, p. 453
26 *Proceedings of the Royal Society B*, 2006, DOI:10.1098/rspb.2006.3480
27 *Animal Behaviour*, 2006, DOI:10.1016/j.anbehav.2005.06.023
28 *Animal Behaviour*, 2004, vol. 67, p. 663
29 *Biology Letters*, 2005, DOI:10.1098/rsbl.2004.0247
30 *Behavioral Ecology and Sociobiology*, 2004, DOI:10.1007/s00265-004-0862-x
31 *Nature*, 2004, vol. 431, p. 646
32 *Journal of Experimental Marine Biology and Ecology*, 2005, vol. 316, p. 29
33 *Animal Reproduction Science*, 2005, vol. 86, p. 1
34 *Journal of Experimental Biology*, 2006, vol. 209, p. 1454
35 *Current Opinion in Neurobiology*, 2003, vol. 13, p. 765
36 *Nature*, 2005, vol. 438, p. 1095
37 *Acta ethologica*, 2005, vol. 8, p. 51
38 *European Journal of Wildlife Research*, 2005, DOI:10.1007/s10344-005-0093-0
39 *Journal of Experimental Biology*, 2006, vol. 209, p. 1454
40 *Biology Letters*, 2006, DOI:10.1098/rsbl.2006.0457
41 *Nature*, 2004, vol. 431, p. 145

Origin of the species

1 www.nature.com/nsu/nibs/200402.html
2 *Nature*, 2004, vol. 431, p. 925 and 2003, vol. 421, p. 335
3 *Science*, 2004, vol. 304, p. 1781
4 *Livestock Production Science*, 2004, DOI: 10.1016/j.livprodsci.2004.11.001
5 *Science*, 2005, vol. 307, p. 1618
6 *Applied Animal Behaviour Science*, 2006, vol. 97, p. 3
7 *Livestock Production Science*, 2004, DOI: 10.1016/j.livprodsci.2004.11.001
8 Ibid.
9 Jeffrey B. Graham, *Air-Breathing Fishes Evolution, Diversity and Adaptation*, 1997, San Diego: Academic Press, 1997
10 California Academy of Sciences
11 *Nature*, 2006, vol. 440, p. 1037
12 *Toxicon*, 2004, vol. 43, p. 471
13 *Nature*, 2001, DOI:10.1038/35070679
14 *Nature*, 2004, vol. 432, p. 94

Colours of the rainbow

1 *Polar Biology*, 2000, vol. 23, p. 147
2 *Biology Letters*, 2006, DOI:10.1098/rsbl.2006.0471
3 *Journal of Experimental Biology*, 2004, vol. 207, p. 2157

4 *Biology Letters*, 2006, DOI:10.1098/rsbl.2006.0440
5 *Vision Research*, 2006, vol. 46, p. 1746
6 *Biology Letters*, 2006, DOI:10.1098/rsbl.2005.0434
7 *Biology Letters*, 2004, DOI:10.1098/rsbl.2004.0227
8 *Proceedings of the Royal Society B*, 2005, DOI:10.1098/rspb.2004.3009
9 *Condor*, 1987, vol. 89, p. 48

Home is where the heart is

1 *Applied Animal Behaviour Science*, 2005, vol. 94, p. 89
2 *Mammal Review*, 2003, vol. 33, p. 174 and *Applied Animal Behaviour Science*, 2005, vol. 90, p. 325
3 *Animal Behaviour*, 1995, vol. 49, p. 827
4 http://news.bbc.co.uk/1/hi/world/africa/3600961.stm
5 *Animal Behaviour*, 2006, vol. 71, p. 457
6 *Journal of Experimental Biology*, 2004, vol. 207, p. 33
7 *New Scientist*, 15 January 2005, p. 42
8 *Animal Behaviour*, 2002, vol. 63, p. 787
9 *Behavioral Ecology and Sociobiology*, 2006, DOI:10.1007/s00265-006-0182-4
10 *Journal of the Acoustical Society*, 2006, vol. 119, p. 627
11 *Behavioral Ecology and Sociobiology*, 2006, DOI:10.1007/s00265-006-0191-3

Weird identity

1 *Conservation Genetics*, 2005, vol. 6, p. 141
2 *Conservation Genetics*, 2005, vol. 6, p. 855
3 *Journal of Experimental Biology*, 2005, vol. 208, p. 2157
4 *Nature*, 2004, vol. 432, p. 913
5 *Nature*, 2005, vol. 433, p. 160
6 *Nature*, 2005, vol. 435, p. 1230
7 *Nature Neuroscience*, 1999, vol. 2, p. 943 and *Nature*, 2006, vol. 440, p. 1190
8 *Nature*, 2002, vol. 418, p. 79 and The Society for Integrative and Comparative Biology annual meeting 2003
9 David Macdonald (ed.), *The New Encyclopaedia of Mammals*, Oxford University Press, 2001, p. 798
10 *Genomics*, 2005, vol. 85, p. 425
11 *Proceedings of the Royal Society B*, 2004, vol. 271, p. 1001

The rarest of the rare

1 *Biological Conservation*, 2006, vol. 129, p. 383
2 *Trends in Ecology and Evolution*, 2002, vol. 17, p. 356
3 *Biological Conservation*, 2000, vol., 96, p. 297
4 www.nature.org.sc
5 www.birdlife.org
6 *Asian Primates*, 2003, vol. 8, p. 8

Death becomes them ...
1 *Primates*, 2006, vol. 47, p. 14
2 *Biology Letters*, 2006, vol. 2, p. 26
3 *The Veterinary Journal*, 2004, DOI:10.1016/j.tvjl.2004.02.007
4 *Nature*, 2005, vol. 437, p. 1107
5 www.birdlife.org
6 Ibid.
7 Ibid.
8 *Biology Letters*, 2006, vol. 2, p. 23
9 *Preventive Veterinary Medicine*, 2006, vol. 74, p. 3

... or not
1 *Raptor Research*, 2004, vol. 38, p. 85
2 *The Journal of the Acoustical Society of America*, 1997, vol. 101, p. 590
3 *Insectes Sociaux*, 2006, vol. 53, p. 32

Resurrection
1 www.birdlife.org
2 news.bbc.co.uk
3 www.tc-biodiversity.org and *Science*, 2006, DOI:10.1126/science.1125631
4 *Nature*, 2003, vol. 426, p. 278
5 *Science*, 2002, vol. 296, p. 1456
6 *PLoS Biology*, 2005, DOI:10.1371/journal.pbio.0030045
7 *Journal of Experimental Biology*, 2006, vol. 209, p. 1575
8 *Nature*, 2004, DOI:10.1038/nature02688